# Florence Nightingale David

Amanda L. Golbeck

# Florence Nightingale David

## A Passionate Probabilist, Statistician, Historian, and Leader

Springer

Amanda L. Golbeck
Little Rock, AR, USA

ISBN 978-3-031-77767-7 ISBN 978-3-031-77768-4 (eBook)
https://doi.org/10.1007/978-3-031-77768-4



This Springer imprint is published by the registered company Springer Nature Switzerland AG
The registered company address is: Gewerbestrasse 11, 6330 Cham, Switzerland

*For my dear partner, Craig Molgaard, who loved the idea of this book, and who was by my side for most of the research and writing.*

# Acknowledgments

We based our work on archival materials from the United Kingdom (University College London Archives, The National Archives, Royal Holloway University Archives) and the United States (University of California—Berkeley Bancroft Library). We made use of publications and recently declassified World War II documents. We conducted site visits and interviews.

Hearty thanks to all who supported the development of this book in the United Kingdom and the United States. Special thanks to Laura Briskman, Rich Levine, Jim Baldwin, Wendy Martinez, Jessica Utts, Dick Greene, Roxy Peck, Emma Styles, Mandy Wise, and Mary Remy. We also wish to acknowledge Colin Mallows, Janet Foster, Rev'd Vivien, Faith Su, Brian Tarran, Dennis K. Pearl, Brian Yandell, Nick Horton, Elyse Gustafson, Janet Foster, Rob Rees, Annabel Valentine, John Wolff, Mario Cortina Borja, Minh Van Nguyen, Hassan Elsalloukh, Mike O'Sullivan, and Joey Lin.

# Introduction

This book is about a British woman in the 1900s who wanted to do head work and persevered to do it within a man's world. Here we reveal what an academic life was like for her, including its ups, downs, and sideways. She was a statistician, but her story can be recognized by any passionate woman with the aspiration to be a professor in any field. We are professors, and her story is our story.

Florence Nightingale (b.1820–d.1910), the world famous nurse, was so passionate about statistics that she considered it "almost a religious exercise [1]." She had many namesakes, one of whom was Professor Florence Nightingale David (b.1909–d.1993), the subject of our book.

Our subject liked to be called David. This is how she was addressed as a professor of probability and statistics in English universities before emigrating to the United States to pursue new academic opportunities [2]. Use of a masculine name surely helped her to fit into the world of academe which was dominated by men.

David may have shared a name and a passion, but there was no one quite like her. She was an original.

The inimitable David is a legend in the field of statistics. She was internationally known, respected, and celebrated. She was a Renaissance statistician [3].

David was a passionate statistician like her namesake, even if she didn't associate statistics with religious exercise. David's passions were broad and extended beyond statistics to also include probability, history, and leadership.

On one of our visits to University College London (UCL) to conduct archival research for our book, we happened upon a display in the UCL Main Building. It was titled *Interrupters and innovators: Journeys in gender equality at UCL*. The exhibition highlighted the accomplishments of some UCL women from a century ago.

The introduction to the exhibit referred to those women who were on display. It stated:

> There were women who were rewarded with professional recognition and personal accolades for their contributions to their discipline, culture and to social reform.

It continued, referring to those like David who were not on display:

> Others, despite equally significant contributions, received much less attention and reward. It falls to later generations to uncover their achievements and restore their reputations.[1]

Our book aims to uncover Professor Florence Nightingale David's achievements and restore her reputation.

As one case study put it:

> ...power structures that have been embedded in our country's premiere centers of historical preservation since their inception helped render [many women's] achievements invisible to future generations. Like a tent stake hammered into the soil at a camground, you could see only the tip of [David's] legacy—and only if you knew where to look. We must rethink how we amass and process information so that the triumphs of women are as easy to trace as those of men.

The above remarks were written about American writer Elsie Robinson, but they apply equally well to David and many other women. To make this point, we have taken the liberty to substitute David's name for Robinson's in the above quote [4].

David's archival materials in the United Kingdom are not contained in a David collection. Rather, the few are scattered into the collections of men, especially Karl Pearson, Egon Pearson, and J.B.S. Haldane. These and other men were the gatekeeprs to her legacy, at a time when it was assumed that the archival materials of women were not worthy to collect and preserve.

We applaud the work of the University of California at Berkeley to remedy this situation for women at Berkeley. We encourage you to visit the 150 Years of Women at Berkeley project and support similar projects in your organizations [5].

David had "enormous energy" and "prolific output," and she was described as "an exciting colleague to work with [6]." As we wrote elsewhere, she "was a prodigy, warrior, professor, writer, leader, and celebrity [3]." Her story deserves our attention.

Particularly noteworthy for us is that David had a powerful will that drove her success. In writing this book, we were inspired by David's determination, and we think you will be too.

## References

1. Cook, E.T.: Life of Florence Nightingale. Macmillan (1913)
2. Reid, C.: Neyman-From Life. Springer-Verlag (1982)
3. Golbeck, A.L. and Molgaard, C.A.: The inimitable F.N. David: A Renaissance statistician. Notices of the American Mathematical Society 68(3), 321–330, March (2021)

[1] Pearlman, N.: Interrupters and innovators: Journeys in gender equality at UCL, exhibition (2019)

4. Gilbert, A.: Opinion: A woman who composed the first draft of history finds herself written out of the history books, CNN (2023) https://www.cnn.com/2023/08/14/opinions/women-history-smithsonian-archives-elsie-robinson-gilbert/index.html Accessed 24 August 2023
5. University of California-Berkeley: 150 Years of Women at Berkeley (2020) https://150w.berkeley.edu/ Accessed 24 August 2023
6. Barton, D.: Obituary: Florence Nightingale David. Journal of the Royal Statistical Society Series A 157(2), 299–301 (1994)

# Contents

# About the Authors

**Amanda L. Golbeck** received her BA from Grinnell College and her MA (anthropology), MA (statistics), and PhD (biostatistics) from the University of California—Berkeley. She is currently a professor emerita in the department of mathematics and statistics at San Diego State University. She is also a professor of biostatistics and the associate dean for academic affairs in the Fay W. Boozman College of Public Health at the University of Arkansas for Medical Sciences. She has had academic positions at San Diego State University (professor of mathematical sciences, statistics division head, and university associate dean for undergraduate studies), Bethany College (provost), Wichita State University (professor of biostatistics and chair of public health sciences), Kansas Board of Regents (vice president for academic affairs), University of Kansas School of Medicine-Wichita (professor of biostatistics and associate dean for research), and University of Montana (professor of biostatistics). Golbeck received a Grinnell College Alumni Award, is an elected fellow of the American Statistical Association, is an elected member of the International Statistical Institute, and was a Fulbright specialist to the University of Latvia-Riga. She received the Committee of Presidents of Statistical Societies Elizabeth L. Scott Award "For her outstanding efforts in enhancing the status of women and minorities, fostering new leadership opportunities for women and men, promoting diversity at all levels, and advocating for a more inclusive, open and supportive atmosphere in statistical sciences." Golbeck is known for her books:

Golbeck, A.L. Ed.: Leadership in Statistics and Data Science: Planning for Inclusive Excellence. Springer (2021)

Golbeck, A.L.: Equivalence: Elizabeth L. Scott at Berkeley. Chapman and Hall/CRC (2017)

Golbeck A.L, Olkin, I. and Gel, Y.R. Eds.: Leadership and Women in Statistics. Chapman and Hall/CRC (2016)

# List of Abbreviations

K Pearson, Karl Pearson
E Pearson, Egon Pearson
UCB, University of California—Berkeley
UCL, University College London
UCR, University of California—Riverside

# Part I

# At the Glass Ceiling

# 1 Deserving (1955–1957)

*...it is still important to remember to teach the young that when you meet a set-back don't sit down and weep, go and get on with your work...*

David, F.N.: COPSS Elizabeth L. Scott Award speech (1992)

## 1.1 Games, Gods and Gambling

Florence Nightingale David was always sad when she finished writing a book. To David, it felt like the death of a close friend. It was hard for her to write that last sentence of the preface, gather up all of her typed pages and put them into a box—this was in 1961, before we had modern desktop computers and the Internet—carry the box to the post office, and send it off to the publisher.

This time, David openly acknowledged her sadness at the end of the preface to her book, *Games, Gods and Gambling* [1]. She wrote, "...once something is written it dies a little and one is always reluctant to call a halt." An American would more likely have written something like "and one is always reluctant to bring writing to a conclusion." But David, perhaps because she was British, chose a more efficient use of language.

Finding a good title for a new book is an art. It can come easily, or it can involve quite a bit of struggle. It is unclear how David landed on the title, *Games, Gods and Gambling*. But it was surely as good as one could get for a book about the history of probability and statistical ideas. How many would not pick up a title that had to do with games, gods, and gambling? One commentator even thought the title was "somewhat mysterious" [2].

David had the ambitious idea for this book about 30 years before she wrote her preface. The book began as a "hobby" project while a student at University College London (UCL), and it spanned her entire career in science. She had been very successful in this career, having achieved the academic rank of reader in the department of applied statistics at UCL. It was the department where the academic discipline of statistics had begun. The only rank higher than reader was that of professor, and that rank was almost always reserved solely for the department head.

By 1955, David already had enough material on the history of dice and gaming to publish an article in the prestigious statistics journal, *Biometrika*, which has published articles on statistical theory and methods since 1901. The journal would have a series on *Studies in the History of Probability and Statistics*. David's study was the first in the series [3]. UCL gave her a grant from the central research


A. L. Golbeck, *Florence Nightingale David*,
https://doi.org/10.1007/978-3-031-77768-4_1

fund to carry out part of the work for the paper which explored the earliest references to random events.

The article would evolve into the first part of David's book. In the article, David explained in detail how dice as we know them evolved from an animal part called an astragalus. We Googled *astragalus* and learned that there is a perennial plant with the same name, a plant that has been formulated into herbal supplements, also with the same name. Not seeing any bones appearing in the preliminary search, we proceeded to Google *astragalus bone* and learned that this bone is known as the *talus bone*, which is an *ankle bone*. People have these bones too, but it was the *astragali* of animals that were used in gaming.

Fortunately, David included pictures of astragali in her article for those of us who do not have a background in anatomy. They don't look anything like dice, we thought. But David convinced us of the evolution, through her masterful study of archaeological materials. Her research even benefitted from her being able to handle ancient dice and critically examine them, thanks to relationships she built with archaeologists who were curators at the British Museum and the Musée de Louvre, as well as with others from UCL who were involved professionally in some way with dice.

The journal article was David's first major step toward the realization of her book. Next, she used the article as the basis of a grant proposal in order to propel her work forward. She especially needed resources to go to where the ancient source materials on the history of probability and statistics were held. In 1955, she applied for and won a research grant from the Leverhulme Trust [4], which found her project to be innovative and original.[1] David's provost sent her a letter of congratulations. An announcement was made in *The Times*.[2] Receiving the grant was a pretty big deal.

David's historical work was raising questions and generating controversies. For example, a staff member at the British Museum's department of Greek and Roman antiquities in London told her he thought she was wrong about the Lydian invention of coinage, and he put her in touch with someone who was interested in the mathematics of the dodecahedron.[3]

The news about David's historical work was already spreading from academic circles to the public. *The Manchester Guardian* published a piece on her work, bemusing the fact that the ancients did not seem to notice that die had an equal chance of falling on any of the faces. Academic publishing was so male dominated that *The Guardian* assumed David was a man and referred to her as Mr. F.N. David.[4] David's colleagues continued to support her work by virtue of their expressions of interest and ideas.[5]

David was asked about Florence Nightingale, who at the time was well known in England but not so well known elsewhere.[6] She said there was no critical appraisal of Nightingale's statistical work, and she was thinking about writing one, but the letters were so scattered that it would be difficult. Was Nightingale a statistician? David thought so, saying: "I would say therefore that she would be the first woman in England to use statistics to draw logical conclusions and as such was I suppose the first woman statistician in England. I can't recollect that there have been any others until we come to the Karl Pearson Era."[7] David was asked about an epigram which read,

[1] Note: Philanthropist William Hesketh Lever became wealthy through the development and international sale of soaps and cleaning products under the company name Lever Brothers. He was given the title Lord Leverhulme, the Hulme part coming from his late wife's maiden name. He died in 1925 and left a portion of his holdings under the name Leverhulme Trust. Today scholars in the United Kingdom can still apply for such a grant up to 500,000 GBP and up to 5 years.

[2] Evans, B.I.: letter to David, F.N. 2 June 1955.

[3] Unkn.: letter to David, F.N. 17 August 1955.

[4] Kendall, M.: letter to David, F.N. 8 August 1955, with *The Manchester Guardian* article attached.

[5] Haldane, J.B.S.: letter to David, F.N. 11 May 1956.

[6] van Dantzig, D.: letter to David, F.N. 24 January 1956. Note: Dantzig was a professor at Mathematisch Centrum in Amsterdam.

[7] David, F.N.: letter to van Dantzig, D. 30 January 1956.

> Upon immortal stairs her shining image waits and not in vain: she holds the lamp of love and courage high to light us home.

Was this written about Nightingale, and who was the author?[8] Unfortunately, David's reply and the name of the author of this epigram remain unknown.

In 1956, David planned to use part of her grant to go to Italy for a month-long pilgrimage. Her idea was to go directly to the historical sources in the great libraries in Italy: Bologna, Florence, Perugia, and Rome. She was looking specifically to access the fifteenth- and sixteenth-century manuscripts on arithmetic. She already knew of five at Bologna, and she thought there were others at the Vatican library in Rome.

David loved libraries and ancient manuscripts. She knew she would find other treasures in the libraries in the process of looking for the materials that she already knew to be there. She wrote ahead to colleagues at universities in the places she planned to visit, asking their help in pointing her to any resources that might be fruitful to her project.[9] She did receive some helpful suggestions that she pledged to follow up on. She had suspicions that the development of mathematics followed a different path than the current thinking, and she was accordingly committed to learn more.[10] When David went to a conference, she would visit local museums to look for clues that would be useful in her history work.[11] David thought her work on the history of probability would extend "for the next 30 years."[12]

David was so committed to her project that she proposed to her department head, E Pearson, that she go to the identified study sites during the Easter vacation. David further proposed to Pearson that she be allowed to take with her the department's part-time librarian, Ms. Jean Edmiston. A professional friend and colleague, Edmiston was already helping David locally in the British Museum and college libraries. She understood and valued the project. David thought that, in Italy, Edmiston could help find relevant materials and arrange with the libraries for copies to be made. She could also help translate the materials, being better at translating medieval Latin than David was. This could reduce the time that David would have to spend in Italy and be away from the department.

Edmiston was willing to go to Italy to help David, even during the Easter vacation. However, David was considerate of others' time and didn't feel it was appropriate for Edmiston to use all of her Easter vacation time to help with the project. She expertly proposed a compromise to E Pearson that Edmiston use 2 weeks of her vacation time and be paid for the other 2 weeks while in Italy. She explained to him: "I think this is a fair request. Miss Edmiston will be getting experience of the way in which Italian libraries are run in addition to helping me with my research and this should make her a more useful member of the college staff."[13]

Pearson supported David's proposal on the basis that the grant wouldn't even be enough to cover all of David's own expenses, much less those of Edmiston. Furthermore, he reasoned, the department was "particularly interested" in David's investigations.[14] As a result, David and Edmiston went to Italy and returned with many treasures for David's book.

---

[8] Skutsck, D.: letter to David, F.N. 3 May (n.d.). Note: D. Skutsck was in the UCL department of Latin.

[9] David, F.N.: letter to Gini, C. 11 January 1956. Note: Corrado Gini was a professor of statistics and sociology at the University of Rome.

[10] David, F.N.: letter to Fortunati, P. 14 March 1956. Note: Paolo Fortunati was a professor at the instituto di Statistica dell'Università di Bologna.

[11] David, F.N.: letter to Statens Historiska Museum director, 4 September 1957. Note: This museum was in Stockholm, Sweden.

[12] Leverhulme Research Awards Secretary: letter to David, F.N. 15 October 1956.

[13] David, F.N.: letter to Pearson, E. 24 February 1956.

[14] Pearson, E.: letter to the University College London secretary, 27 February 1956.

With the Levenhulme Trust funding, David went to Bologne, Firenze, Rome, and then Geneva. She had an accident while there and broke a bone in her leg. As she put it, this curtailed her wanderings by having "to carry what felt like a ton of plaster about with me."[15] David was very grateful for the grant and said: "The possession of the grant enabled me to make many happy and fruitful acquaintances in Italy all of whom offered to help me when I go back which will expedite the work next time. Through having the grant also I have become aware of the amount there is to do and how little I have done. I estimate that my vacations for the next 30 years will be fully occupied."[16]

Over the 30 years she worked on the book, David made many friends in the profession, a "formidable" who's who list of top scientists who you will meet as you continue to read David's story. These friends contributed many ideas for David's book *Games, Gods and Gambling*. When David got a new idea for her book, she would write it by hand into a notebook. Eventually she filled numerous notebooks, in fact "many each year." Staying with a project for 30 years took an enduring love for the subject matter and a great deal of perseverance, to be sure.

One friend that stood out to David as she wrote that last sentence of her preface was Sir Maurice Kendall. He was chairperson of statistics at the London School of Economics, and his office was less than a mile away from David's. Kendall and David had both done significant research on measures of relationship between things, which statisticians call correlation. David dedicated her book on the history of probability to Kendall "because of his continuing interest and encouragement," helpful ideas, and "conscious prodding."

David had already written some statistical sciences books, and she would eventually write some more. But this book was different. *Games, Gods and Gambling* showcased David's breadth as a scholar. It showed that she wasn't just a statistician; she was a *liberal artist* with quantitative leanings. She wove archaeology, biography, classics, fiction, and poetry into her history of probability. She added a full dose of human interest into an otherwise technical subject.

Sometime in the first or second grades in the United States, we are taught basic mathematical symbols, like the plus, minus, and equal signs. But do you know where these signs came from? David told us in her book:

> *During the fifteenth and sixteenth centuries the Hindu method of number-writing spread through Europe and across the Channel to England. Also, as might have been expected, algebraic symbolism improved, the signs plus (+) and minus (-) making their first appearance ... Robert Recorde (1510-58) uses the plus and minus symbolism in his Grounde of Artes published in 1540. Seventeen years later in 1557 he published the Whetstone of Witte in which, for the first known time, the sign for equality appears. He chose it, he endearingly remarks, "because noe 2 thynges can be more equalle than a pair of parallels."* [1] (p. 61)

This sample of writing shows how David combined scholarship and storytelling, and she did so throughout her book. David didn't just tell *what* happened; she contextualized it with the *who* and the *why*.

*Games, Gods and Gambling* was a revolutionary accomplishment. Only one other book on the history of probability had ever been written. It was Isaac Todhunter's 1865 book, *A History of the Mathematical Theory of Probability: From the Time of Pascal to That of Laplace* [5]. David had great respect for Todhunter's book. She found it to be mostly correct, explicit, informed, fair, and precise.

Todhunter's book began with Pascal and continued to Laplace in the early nineteenth century. David's book began earlier with ancient Egypt prehistory. She originally intended her book to go to the present day, but she decided to end with de Moivre in the mid-eighteenth century, leaving the rest for a possible future book. David thought Todhunter's book not only started too late in terms of chronology, but it ignored important narrower threads of the history.

---

[15] David, F.N.: letter to Leverhulme research awards secretary, 12 October 1956.

[16] David, F.N.: handwritten letter to the Leverhulme Trust, n.d.

David had the temerity to break the 100 years of silence on the subject of the history of probability which followed Todhunter's culturally important book. While she admired Todhunter's book, she found it to be dry, tedious, and austere. Hers was considered to be a readable masterpiece in the history of probability. One could say that *Games, Gods and Gambling* brought the history of probability... not just *back* to life, but *more* to life.

David's *Games, Gods and Gambling* was meant to be evidence-based. She aimed to create a scholarly text that could be read for both "profit" because she had checked the original mathematical documents and for "pleasure" because she had checked the stories being told about "great men."

David saw history of science writing as a continuum. At one end of the spectrum, there are writers who focus to an extreme on technical facts. They "write severely on the subject-material and leave the men who created it as wooden puppets"... David found Todhunter and other theorists to fall at this end. Their work is difficult to read for pleasure. At the other end of the spectrum, there are writers who tell "interesting (and often apocryphal) stories about personalities, with no real attempt to place their achievements." Their work doesn't provide us with understandings of theory or contextualizations of accomplishments. Such storytellers, argued David, leave us with little understanding of theory.

In 1961, David wrote into her preface that she wanted to write a book that fell into the middle of this spectrum. As she put it:

> *The man creates the mathematical theorem, but the events of a man's life create the man, and the three are indissoluble. Thus I would hold, for example, that it is not enough to remark on and wonder at de Moivre's analytic genius, but that one should also realize that poverty was his spur, that much of his work might not have been achieved had he been sure of a post which would have brought him some leisure and about which he wrote so longingly to John Bernoulli.* [1] (p. x)

Women in science were so scarce across this part of history that even David wrote about "the man." And after all, the history was all about men. David names them in chapter titles: Tartaglia, Cardano, Galileo, Fermat, Pascal, Huygens, Wallis, Newton, Pepys, Bernoulli, de Montmort, de Moivre, Buckley, de Coste, and Carcavi. David calls these men *masters* and devotes about a third of her book to translations of papers or letters written by or between them. David's aim was to evaluate the contributions of these and other men to the history of probability.

On the pages before her preface, David included excerpts from a poem written by James Elroy Flecker in 1913. The poem, titled "The Golden Journey to Samarkand", used to be very popular in England. Samarkand is in Uzbekistan. There are many ancient mausoleums and mosques there, and it is part of the ancient trading route between China and the Mediterranean known as the Silk Road. David included five stanzas. Here we will repeat two:

> *What shall we tell you? Tales, marvelous tales*
> *Of ships and stars and isles where good men rest,*
> *Where nevermore the rose of sunset pales,*
> *And winds and shadows fall toward the West.*
> ...
> *And how beguile you? Death hath no repose*
> *Warmer and deeper than that Orient sand*
> *Which hides the beauty and bright faith of those*
> *Who make the Golden Journey to Samarkand.* [1] (p. vii)

David thought that arithmetic and probability spread down the Silk Road from China. Consequently, she always wanted to go down that road to explore the monasteries where she believed the learning had been centered [6].

One commentator at Cambridge University remarked that this poem "is a wonderful evocation of the hopes, potential mysteries and unforeseeable revelations promised in the open-ended adventure... and of course I loved the richness and fluidity of the language but also the way it doesn't take itself too seriously" [7]. It is interesting that this commentator was, like David, a mathematical scientist.

What meaning did David find in this poem? We can speculate that David's book had been a similar open-ended adventure for her, full of hopes, mysteries, and revelations. Her journey took her, and now takes us as readers, into far away and even exotic places. Her prose was rich, fluid, and unassuming. In addition, we find the tone of the poem to be melancholy. It mirrors David's sadness at completing her book, a sadness that she articulated so clearly in her preface.

## 1.2 Hitting the Sweet Spot

David did accomplish her goal of writing a book that was accessible, excellent scholarship, and a compelling read. Her *Games, Gods and Gambling* book was published by Charles Griffin and Co. Ltd in London in 1962. In terms of being accessible, David's book cost 6.50 USD, which is under 60 USD in today's currency. In addition, it was smaller and more readable than Todhunter's book. It had something to offer for both the professional and the average reader who has no knowledge of probability theory. As one reviewer put it, "This book will be valued by all persons interested in the history of probability... She has done this job [supplementing and expanding on Todhunter's work] well" [8].

In terms of excellent scholarship and a compelling read, the book received many glowing reviews after it was published. Reviews in professional journals called the book "an eminently readable blend," "fascinating reading," "entertaining, yet serious," full of "interesting facts and fresh ideas," and "very useful and readable." One reviewer put out an urgent plea for more books similar to David's that would cover other time periods of the history of probability and its applications [2]. Another expressed gratitude that David didn't wait to include material past de Moivre and looked forward to her future historical writings [8].

Although the reviews were mostly glowing, there were a few critical remarks. One reviewer accused David of making moral judgments and of watering down some of the mathematical derivations. He felt that David missed the sweet spot previously described between the biography and the technical. He commented: "In consequence, there is some risk that the casual reader will not appreciate the difficulty and depth of many of the problems." However, then the reviewer cut some slack: "But no doubt Professor David has reserved such points for a monograph which she is planning..." [9]

Another reviewer thought the book had a general vagueness and that David should have conducted a deeper analysis [10]. A different reviewer gave the book a four (out of five) star review, saying that it "... is written in an attractive, narrative style that seems a bit old-fashioned and opinionated at times but never monotonous. Her facts are well documented and her viewpoints are mostly well argued, yet she does not attempt an exhaustive or mathematical treatment. Therefore the book remains very readable and stimulating until the end" [11].

One more reviewer, who was a historian and not a mathematical scientist, thought David could have carried the main story "a step further or to present additional source material on the probability theory proper rather than general background information... She has to some extent neglected to show us the inherent continuity (or discontinuity) of the mathematical ideas as they developed through the centuries." This reviewer acknowledged the difficulty of writing this kind of book: "A comparison of the achievement of great men is a difficult and dangerous task: too many different factors combine to form a man's intellect and personality. Some readers may disagree with the author in her rating[s]..." [12].

Yet another reviewer thought that David did hit the sweet spot. He described the book as "... by no means as dry as a scholarly historical text nor as inaccurate as a truly popular account. The authoress says she has tried to steer a midway course and she has done so quite successfully... it is surprising to see no mention whatsoever of the growth of probability as a measure of degree of belief rather than as a concept associated with repetitions. Are there no references in literature to this idea?... I enjoyed this book and you will too. It is good bedtime reading for a statistician" [13]. This reviewer was Dennis Lindley, who we will encounter later in David's story.

Here we just note that in the present day, the word *authoress* is dated and objectionable. Its use would likely not have pleased the very accomplished David, even back then. After all, she published under the name F.N. David, intentionally refraining to use her feminine first and middle names. Nor would the reference to her career-long work as bedtime reading have pleased her. She intended the book to be more serious than that. Lindley's surprise to see David treat probability as a concept associated with repetitions (*frequentist*) rather than a measure of degree of belief (*Bayesian*) is really no surprise at all. It sums up the philosophical differences between the two statisticians. As will become clear later in our story, this wasn't an innocuous philosophical difference. Rather, it was one that would ultimately help to change the course of David's career.

David's close colleague and former student D.E. Barton wrote this in his obituary of her: "[The book's] merit consisted of the real insights that it provided into the development of the primitive notions of mathematical probability. The informal presentation was a fair reflection of the confusion that was prevalent at the birth of the subject (even in the minds of such giants as Galileo, who laid the first firm foundations)" [14].

There have been multiple editions of David's book. We own the 1998 Dover Publications, Inc. edition. The back cover declares the book to be "A really delightful, erudite, and completely charming account of the history of probability and chance." The charm can be made evident via a few examples. These examples are from David's Chap. 10: "Bills of Mortality". This chapter had special significance for her, because as a girl she aspired to be an actuary, only to be told that she could not because she was a woman.

A first example has to do with the lack of interest in probabilities on the part of the English.

> *Modern statistical method and practice has its twin roots in the calculus of probabilities and in the collection of numerical data. It would be outside the main theme of this present treatise to trace the rise of actuarial science and the growth of vital statistics, but at this juncture, when the calculus of probabilities was well launched, it is interesting to ask what the English were doing. Pre-eminence in mathematics had passed from Italy to France and was shortly to pass to England with the advent of Newton. Speculations and research into probability theory did not, apparently, interest mathematicians across the Channel. The English, as usual, were preoccupied with concrete facts.* [1] (p. 98)

A second example has to do with John Graunt, who because of his pioneering work with death and birth records is recognized as the first English vital statistician.

> *John Graunt's manuscript was presented to the newly formed Royal Society in 1661 and he was elected a Fellow in 1662, presumably on the strength of it. Stories (probably apocryphal) are told that difficulties were raised about his election because he was just a shopkeeper and that Charles II recommended him personally to the Society, adding moreover that if they found any more such shopkeepers they should also admit them. Perhaps the fulsome references were not lost on the king, although this would imply that he had read the book.* [1] (p. 102)

A third example preserves the amazing quote from Florence Nightingale (b.1820–d.1910), after whom David was named, about the essence of statistics.

> *It would be agreed that the risks inherent in the processes of birth, marriage and death can only be calculated for a short forward period, and then only provided the structure of the population at risk is not changing.*
>
> ...
>
> *The earlier workers in the demographic field - and their successors over several centuries - believed that this stability was a direct manifestation of the purpose of God. For instance, Florence Nightingale, after some lengthy*

*calculations, wrote: "The true foundation of theology is to ascertain the character of God. It is by the aid of Statistics that law in the social sphere can be ascertained and codified, and certain aspects of the character of God thereby revealed. The study of statistics is thus a religious service."*

Then David wrote this about her namesake, on the same page in a footnote: "With truth indeed did she call herself *a passionate statistician*" [1] (p. 103).

There was nothing shy or timid about David. She chose for the dust jacket of her book a sketch of the Greek goddess of chance, Tyche. The goddess was half-naked.

Society was more puritanical then. When David tried to carry a copy of the book into the United States, the customs officer required her to dispose of the dust jacket [15]. One can imagine her going through today's TSA screening, tossing the dust jacket into the huge waste can along with the too-large (100 ml or more) bottles of liquid that she forgot to leave at home.

Our paperback copy of the book looks very different. It isn't entire clear to us what the revised cover is meant to depict. Rippling water, perhaps? In any case, there is no scantily-clad Tyche. Problem solved.

Current booksellers call David's book an "absorbing history" with "fascinating allusions" that give it "universal appeal." You can still buy a new copy on Amazon, but you will have to pay 599 USD for a hard cover copy of the book.

## 1.3 Her Claim to the Title

Being a faculty member in the department of statistics at UCL was David's life. Back in 1957, when she was 48 years old and in the thick of writing *Games, Gods and Gambling*, David took a moment and counted the many years since she first arrived at UCL when she was 19 years old. She was a student for 6 years, a lecturer for 15 years (some with the military during World War II), and a reader (one rank lower than professor) for 8 years. She added up the years and figured she had been at UCL for 29 years. She definitely wasn't a job hopper. E Pearson had been David's supervisor for almost all of those years. He had advanced her career progression and promoted her up the academic ladder.

David invested countless hours on her work—during the week, weekends, and vacations. She made a formal request to do some of her work at home rather than doing all of it in her campus office. E Pearson explained the need: "Strictly speaking, she could work for 365 days of the year in her room in College, but as every one knows this is not the way which original mathematical type of work can be done, and I would regard the doing of vacation-time research [away from her university office] as right and proper."[17] Thus, it became "an indispensible condition of the appointment of Dr. Florence N. David that he [sic] should perform part of the duties of his [sic] appointment at home...."[18]

Working diligently all those hours and days, David built an impressive scholarly record. She had every right to be extremely proud of it. Faculty members work hard to build a CV that will impress their colleagues and support their claims to prominence in their field.[19] By 1957, David already had 4 books and 43 scholarly publications on her CV. She also had a large number of classified reports that she prepared for her government during World War II. At age 48, she had an impressive CV. As

[17] Pearson, E.: letter to Gueterbock, 24 February 1955.

[18] Tax Assessment—Study Accommodation four documents for years 1949–1962. Note: The last such form, dated 26 April 1961, changed the gendered language.

[19] Note: A curriculum vitae, or CV, is somewhat like a resume, in that it contains a summary of education and experience, but it is generally longer because it contains the itemization of things like scholarly publications which provide important evidence of such prominence.

she put it many years later, "I had many more publications than anybody who was likely to be asked to take over" for her department head E Pearson, who was expected to be retiring at the end of the decade [6].

It is no wonder that David hoped, and even expected, to someday be promoted to the rank of professor. This would give her a raise (from her current 2,155 GBP[20]) and allow her name to be printed on the top of the department stationery. More importantly, since the professor title and department head responsibilities would normally come in the same package, she would have the opportunity to lead the vision and set the agenda for her discipline in her university. In the United States, we would say that she was ready to be promoted to the position of professor and department chair. In the United Kingdom, we would say that she had claims to the title of professor.

David enjoyed excellent performance evaluations from E Pearson as he promoted her up the ranks. An early evaluation, for her appointment to lecturer, said: "She has without question proved herself well qualified as a teacher... She has always been a hard worker... I think her powers of independent thought are developing... She has established herself as a really useful member of the staff of the Department...."[21] A mid-career evaluation, for her promotion to reader, said: "Her original work in the field [of Probability Theory] has both individual character and high quality... She is an excellent teacher to all types of students... [She has] power of handling practical statistical problems...."

Importantly, in the same mid-career evaluation, E Pearson gave David credit for helping to sustain a quality work environment within the department: "The growth of this Department is the result of discussion and experiment in which she has played a most essential part. Many of the ideas which have been incorporated were hers... The Department of Statistics, under one title or another, has had a long tradition of good workmanship and 'happy' organization. Dr. David's activities within the Department are such as to continue and strengthen this tradition...."[22]

Then only a few years later, E Pearson supported David's transfer to a faculty line previously established as a reader in statistics in 1935 for the renowned statistician Jerzy Neyman. This reader position was created "... to introduce into the Department an element that had hitherto been lacking—namely to give full and permanent representation to teaching and research on the Mathematical Theory of Probability." Neyman held this readership for 3 years before being recruited to the University of California - Berkeley (UCB) in the United States.

Pearson had held this reader line open for about 20 years because he felt there was no suitable candidate.[23] Perhaps in all those years, he was looking for and could not find someone of sufficient caliber to replace Neyman. David had to feel deeply honored when E Pearson transferred her to the position.

Neyman and E Pearson were close. Together they had developed a new paradigm for statistical hypothesis testing that became widely adopted in the field. A 50-year retrospective summed it up: "The Neyman-Pearson theory makes two principal recommendations: Evaluate the behavior of statistical procedures in terms of some performance criterion, and then opt for the procedure that performs best. The application of these recommendations requires a probabilistic model for the observations, although such models can be much broader than those originally envisaged by Neyman and Pearson" [16]. Everyone who takes an introduction to statistics class learns something about the Neyman-Pearson theory.

When the position was transferred to David, it involved a change in title from reader in statistics (as held by Neyman) to reader in probability and statistics (as would be held by David). E Pearson wrote:

[20] Finance Secretary and Accountant: letter to David, 14 August 1955.

[21] Pearson, E.: statement regarding the promotion of David, n.d.

[22] Pearson, E.: statement re David's proposed promotion to the position of reader, 23 February 1949.

[23] Pearson, E.: note on the established readership in statistics, estimates 1955–1956 (n.d.).

"Since Dr. Neyman left she [David] has carried on and developed the teaching of probability theory in the Department and much of her research has been in this field; she has also for long been interested in the History of Probability... There is a strong feeling in statistical circles that lecturers on the History of Probability, discontinued since K Pearson lectured on the subject during the years 1921-33, should again be given in the University. Dr. David is admirably suited to start such a course and her transfer to a readership in Probability and Statistics would make it clear that University College was prepared to take its rightful lead in this matter."[24] Thus, David became the first person in the UCL department of statistics to hold a title in both statistics and probability.

David had a strong claim to the title of professor. E Pearson had declared her teaching to be excellent and her research to be high quality. David had earned a DSc degree in 1952, which in the United Kingdom is a higher degree than the PhD. She had been put into Neyman's old position and in December 1955 became the department's first reader in probability and statistics.

David had a strong reputation in the community and the university. In 1955, she was appointed to a 3-year term representing the UCL senate on the governing board of the Camden School for Girls in London.[25] When she was again appointed to the UCL professorial board in 1957, the UCL secretary to the senate wrote to her: "I myself am worried about our weakness on the Senate, and you are of very great assistance to us there."[26] In 1956, she supported the appointments of William Henry Kruskal (b.1919–d.2005) and Elizabeth Leonard Scott (b.1917–d.1988) to the Royal Statistical Society.[27]

Also supporting her claim, David had an international reputation. She was elected to be a fellow of the Institute of Mathematical Statistics (IMS) in 1946, having "demonstrated distinction in research in statistics or probability, by publication of independent work of merit." She was the third woman to be so elected, following Hilda Geiringer in 1940 and Gertrude M. Cox in 1944 [17]. David was elected in 1951 to the International Statistical Institute (ISI), only the third woman ever to be so elected.[28] Additionally, she was elected to be a fellow of the American Statistical Association (ASA) in 1954.[29] According to the American Statistical Association web site, "The designation of ASA Fellow has been a significant honor for nearly 100 years. Under ASA bylaws, the Committee on Fellows can elect up to one-third of one percent of the total association membership as fellows each year. [para] To be selected, nominees must have an established reputation and have made outstanding contributions to statistical science. The Committee on Fellows evaluates each candidate's contributions to the advancement of statistical science..." [18]. Being elected to all three honorary titles (IMS fellow, ISI elected member, and ASA fellow) put David into an exclusive category among her peers.

David was sent a letter informing her she had been appointed to be UCL reader in probability and statistics in 1955. Her appointment would span from 1 October 1955 through the session in which she became age 60.[30]

In 1957, E Pearson was 62 years old and approaching retirement. As a young man, he was reportedly introverted and had feelings of inferiority. His father, K Pearson, had been the founding head of the applied statistics department and has been called the "father of mathematical statistics." E Pearson had inherited the applied statistics part of his father's position and was aware that he had "in

---

[24] Pearson, E.: memorandum to the board of studies in statistics, estimates 1955–1956, 18 February 1955.

[25] Phillips, J. Hood: letter to David, F.N. 17 November 1955.

[26] Gueterbock, E.A.L.: letter to David, F.N. 30 January 1956.

[27] Royal Statistical Society Nomination Paper numbers 8153 and 8156.

[28] Pearson, E.: statistics department estimates 1952–1953, David, 28 February 1952.

[29] David, F.N. List of appointments, awards, and publications since conferment of the title of reader in statistics in 1955.

[30] The UCL Principal: letter to David, F.N. 28 November 1955.

so many ways throughout life" seemed to have "got there" because he was the son of K Pearson.[31] The younger Pearson was highly protective of his father's reputation, and he saw the statistics chair at UCL to be "the premier statistics chair in Britain" [19].

E Pearson was starting to think about the next few years and his successor. No woman, except for one, had ever been given the title of professor at UCL. Given the times, it would have taken someone who felt exceptionally secure in their own legacy to welcome a woman to succeed him.

His father had helped him to succeed to the professor and department head position, so E Pearson thought he could exert the same influence and name his own successor. In 1957, a few years away from retirement, E Pearson wrote a letter to his provost. He marked the letter confidential.

E Pearson communicated with the provost about the possibility of conferment of the title of professor on Dr. F.N. David. He wrote: "... the suggestion has been put to me and supported by a number of persons, and there is without doubt a case for consideration. Dr. David's publications are extensive and are continuing without any signs of flagging; some, if not all, of them are of very good quality. She has been responsible for the supervision of the work of more than half the statistical post-graduates during the past 10 years, and she has been closely concerned with the development and expansion of our own degree courses since the war."[32]

But then in just two sentences, E Pearson sealed David's fate: "On the other hand I am inclined to think that she is not by temperament the right person—or the best person—to become head of what is really the senior Statistics Department in the country."[33]

By her own estimation, David worked twice as hard as the men around her. Some 30 years later, she explained: "Well, in my day you had to do 200 percent better than the nearest man they could possibly give the job to before you got it" [6]. Sadly, E Pearson went to some lengths to arrange for his position to go to another man.

Later on, in 1992, the Committee of Presidents of Statistical Societies gave David the first Elizabeth L. Scott Award. The Committee of Presidents of Statistical Societies is a consortium of presidents and executive directors of five major statistics associations: the American Statistical Association, Eastern North American Region of the International Biometric Society, Institute of Mathematical Statistics, Statistical Society of Canada, and Western North American Region of the International Biometric Society [20].

David had this to say at the award ceremony:

> *...I am honored to have been chosen as the recipient of the Elizabeth Scott Award. I go back a long way in time. I started my academic life in 1929 and wanted to be an actuary. So in 1931, I visited an insurance firm to ask about applying for a career fellowship. I was turned down and when I asked why I was told that although I was the best qualified person applying I was a woman. I went home raging to my father but all he said was "You will meet this all your life. Go away and get on with your work." Fortunately a week or two later I went to see K Pearson who was encouraging, giving me a place in his laboratory and seeing that my college scholarship was extended. And so I learnt the two lessons which have haunted me all my life. First that women tend to be unfairly treated and second that if you persevere difficulties can be overcome. Now, due in the main to the opportunities of the 1939-45 war, conditions are better for women but it is still important to remember to teach the young that when you meet a set-back don't sit down and weep, go and get on with your work....*

David met the setback given to her by E Pearson. Fortunately for all of us, she went on with her work and became a role model for the profession.

---

[31] Pearson, E.: handwritten letter to Neyman, J. 27 March 1966.

[32] Pearson, E.: letter to University College London provost, 29 November 1957.

[33] Pearson, E.: letter to University College London provost, 29 November 1957.

## References


1. David, F.N.: Games, Gods and Gambling: The Origins and History of Probability and Statistical Ideas from the Earliest Times to the Newtonian Era. Griffin, London and Hafner, New York (1962)
2. Fisz, M.: Review: Games, Gods and Gambling: The Origins and History of Probability and Statistical Ideas from the Earliest Times to the Newtonian Era, by F.N. David. Econometrica 32(1/2), 236 (1964)
3. David, F.N.: Studies in the history of probability and statistics I. Dieing and gaming (A note on the history of probability). Biometrika 42(1/2), 1–15 (1955)
4. Leverhulme Trust: Leverhulme Trust (2021) leverhulme.ac.uk (2021). Accessed 31 July 2021
5. Todhunter, I.: A History of the Mathematical Theory of Probability from the Time of Pascal to That of Laplace. MacMillan, Cambridge and London (1865)
6. Laird, N.M.: A conversation with F.N. David. Statistical Science 4(3), 235–246 (1989)
7. Forster, T.E.: Comments on The Golden Journey to Samarkand (1913) https://www.dpmms.cam.ac.uk/~tf/poem36.html (2021). Accessed 6 August 2021
8. Thomas, G.B., Jr: Review: Games, Gods and Gambling: The Origins and History of Probability and Statistical Ideas from the Earliest Times to the Newtonian Era, by F.N. David. The American Mathematical Monthly 71(4), 452–453 (1964)
9. Thatcher, A.R.: Review: Games, Gods and Gambling: The Origins and History of Probability and Statistical Ideas from the Earliest Times to the Newtonian Era, by F.N. David. Journal of the Royal Statistical Society Series A (General) 126(2), 319–320 (1963)
10. Ore, O.: Review: Games, Gods and Gambling: The Origins and History of Probability and Statistical Ideas from the Earliest Times to the Newtonian Era, by F.N. David. The Annals of Mathematical Statistics 35(1), 447 (1964)
11. Crombrugghe, D.D.: Review: Games, Gods and Gambling: The Origins and History of Probability and Statistical Ideas from the Earliest Times to the Newtonian Era, by F.N. David (2001) https://www.amazon.com/Games-Gods-Gambling-Probability-Statistical/dp/0486400239 Accessed 29 July 2021
12. Scriba, C.J.: Review: Games, Gods and Gambling: The Origins and History of Probability and Statistical Ideas from the Earliest Times to the Newtonian Era, by F.N. David. Isis 55(2), 216–218 (1964)
13. Lindley, D.V.: Review: Games, Gods and Gambling: The Origins and History of Probability and Statistical Ideas from the Earliest Times to the Newtonian Era, by F.N. David. Journal of the Royal Statistical Society Series D (The Statistician) 13(2), 158–159 (1963)
14. Barton, D.: Obituary: Florence Nightingale David. Journal of the Royal Statistical Society Series A (Statistics in Society) 157(2), 299–301 (1994)
15. The Associated Press: Dr. Florence N. David, statistician, dies at 83. The New York Times, Wednesday, 18 July 1993 (1993)
16. Lehmann, E.L.: The Neyman-Pearson theory after fifty years. In: Le Cam, L.M., Olshen R.A., Cheng, C., Neyman, J., and Kiefer, J. (eds.) Proceedings of the Berkeley Conference in Honor of Jerzy Neyman and Jack Kiefer, Volume 1, Wadsworth, California 1047–1060 (1985)
17. Institute of Mathematical Statistics: Nominations for IMS Fellow. (2021) https://imstat.org/honored-ims-fellows/nominations-for-ims-fellow/. Accessed 7 August 2021
18. American Statistical Association: ASA Fellows (2021) https://www.amstat.org/ASA/Your-Career/Awards/ASA-Fellows.aspx. Accessed 31 July 2021
19. Reid, C.: Neyman - From Life. Springer-Verlag, New York (1982)
20. Committee of Presidents of Statistical Societies: Elizabeth L. Scott Award (2021) https://community.amstat.org/copss/awards/scott. Accessed 7 June 2021

# Part II

# Up the Hill

# 2 Growing (1909–1928)

*A girl well above average in every way... she has proved herself an excellent leader & [is] a girl of strong personality... I feel she will have an excellent influence on many with whom she comes in contact.*

McKay-Ohm, D.: letter of recommendation for David, F.N. (1928)

## 2.1 Storytellers, Mysteries, and Books

When David was born in 1909, her parents named her Florence Nightingale David. *The* Florence Nightingale was awarded the Order of Merit 2 years earlier. This was a monumental prize, given to at most 24 living people in the Commonwealth lands for their accomplishments in the military, sciences, arts, literature, or promotion of culture. The award has been described as "quite possibly, the most prestigious honour one can receive on planet Earth" [1] and the "pinnacle" of the British system of honors [2]. Nightingale was the first woman to receive it [3]. David's parents, being educators, undoubtedly took notice.

Nightingale was known as "the lady with the lamp" for her nursing of wounded British soldiers during the Crimean War with Russia (1853–1856). She was a national heroine for this work, and a leader of many social and public health causes in Great Britain.

Of special significance to our story about David, who was a statistician, is that Florence Nightingale was a passionate statistician in addition to being a nurse. She "used her passion for statistics to save lives of soldiers during the Crimean war." She showed that deaths of English soldiers during that war were mainly due to disease, not trauma or wounds. This was a new idea for military medicine of that era. She also did "groundbreaking work in data visualization that continues to be influential to this day." She was best known for her beautiful "coxcomb" designs, which are like pie charts but more intricate. Indeed, Nightingale was both "the lady with the lamp" and "the lady with the data" [4].

Florence Nightingale is a lot for a little girl to live up to, whether aspiring to be a nurse or statistician or whatever. We found a blog of the *American Journal of Nursing* where the author articulates ten lessons to be learned from Nightingale: never stop learning; ground your work in evidence; follow your convictions; treat every person holistically; know your strengths and weaknesses; use your network to accomplish more than you can on your own; speak and write often



A. L. Golbeck, *Florence Nightingale David*,
https://doi.org/10.1007/978-3-031-77768-4_2

to share lessons and learnings; change what you perceive needs to be changed; don't blame others for things you are not willing to try to change yourself; and keep to high standards [5].

High principles such as these are why Jerzy Neyman thought David's parents "should be shot" for naming her after *the* Florence Nightingale [6] (p. 132.). But David did live up to all of these principles, and more, as we shall reveal in the telling of her life story.

We don't know for sure why David's parents named her after the lady with the lamp. One idea, as promulgated in several online biosketches, is that David's parents and Florence Nightingale were friends [7].[1] A popular book additionally states that the lady with the lamp used to retreat to the sanctuary of the David home in the West Midlands County of Great Britain in order to rest and rejuvenate [8] (p. 152). These promulgations are not, however, supported by primary sources, nor do they fit the chronology of their main characters' lives.

Another idea is that David and *the* Florence Nightingale were relatives. This is based on a vague, distant memory of the graduate student days at Berkeley of Craig Molgaard. According to the memory, David professed to a class at Berkeley in the early 1980s that she was Florence Nightingale's niece. Wearing blue jeans, a cowboy shirt, and boots, David talked to the class about Florence Nightingale's last, bedridden days as if living together in the same house. David said the lady with the lamp treated her caretakers badly, implying very heavily she knew Nightingale and was among the poorly treated caretakers, even saying Nightingale was her aunt. She laid it on thick. Stories such as these were passed around to other graduate students beyond the class. Maybe there was a blood relation, and it has been lost in the ancestral records. Or maybe David was sometimes prone to exaggeration while teaching. All that we can say for sure is that David was a great teacher and could be a magnificent storyteller.

Ancestry research failed to turn up a family tree with *the* Florence Nightingale in it. The research did indicate that one of David's ancestors had the first name *Florence* and middle name *Nightingale,* but her last name was *Appleton.* So David did have a grandmother Florence Nightingale Appleton who could have been mistaken for *the* Florence Nightingale.[2]

While researching this book, we were connected to one of David's relatives, Anne.[3] Anne also could not find an ancestral link between the two families. David was Anne's great aunt (Anne's grandmother's sister). Anne wrote: "I was always told that [Florence Nightingale] was related to [my family] and that the lace head veil that I have belonged to her. But after doing my family tree I figured that maybe she was just a family friend as I couldn't find the link."[4]

There are other stories that David reportedly told in her lifetime, indicating we probably will never know for sure where the truth lies. For example, David reportedly once stated she was statistician George Box's (b.1919–d.2013) doctoral advisor [9]. Box, at age 90, on the other hand claimed three men as his examiners.[5]

David had a number of doctoral students while she was at UCL who became well known in the field of statistics. One was Colin Mallows.[6] When we asked him about Box, Mallows pointed out: "In an obit of Box, I read that at UCL he was a student of E Pearson [the head of the statistics department]. Well, so was I. I think it was policy to list everyone seeking a higher degree as Pearson's student."[7]

---

[1] Note: This is one of the several examples.

[2] Note: Jessica Utts, professor emerita of statistics at the University of California, Irvine, constructed David's family tree using ancestry.com.

[3] Note: The connection to Anne was graciously made by Utts.

[4] [Last name unknown], Anne: email to Jessica Utts, 22 September 2021.

[5] Note: Box named examiners Pearson, E. (b.1895–d.1980), Herman O. Hartley (b.1912–d.1980), and Maurice S. Bartlett (b.1910–d.2002) [10].

[6] Note: Mallows is known for Mallows′ $C_p$. This is a measure of how well a model fits in ordinary least squares regression.

[7] Mallows, C.: email to Golbeck, A.L. 16 September 2018.

Going back to why David's parents named her Florence Nightingale, it is also possible that David's parents gave her the name out of deep admiration for her namesake. We think this is the most likely reason. Who wouldn't want their daughter to be like Florence Nightingale? And it didn't hurt that David's mother's and maternal grandmother's first names also were *Florence*.[8]

When she was a little girl, David's closest mentors were storytellers. Stories, told well, have the power to awaken and stir our minds and imaginations. Good teachers and ministers accordingly are professional storytellers. David would have had her mind and imagination awakened and stirred by these professionals, and she would have learned a lot about storytelling from them. This would especially be the case if, as it were, the teachers lived with her and were her parents, and if the minister was at the church right across the street from her house and was her tutor.

We wanted to experience the village of Ivington[9] in the United Kingdom where David heard her childhood stories. Our interest was piqued by Ivington being so old that it is mentioned in the 1086 Domesday Book. This book is King William the Conqueror's awesome, detailed, and unalterable statistical survey and census of many parts of England and Wales. Therein, Ivington was recorded as "Ivintude," which is an Old English name for "estate associated with a man called Ifa" [11] (p. 198). That David was from a place named in the Domesday Book indicates how deep her British roots were. At the end of her life, David would have her body returned, from where she was then living in the United States, to this part of Britain to be buried.

Because we weren't satisfied with just reading about Ivington on the web, we journeyed from the United States to the United Kingdom. We took a 2 h 35 m train from London Paddington to Hereford, transferring in Newport, South Wales. It was pleasant to see the British countryside while anticipating what we would see and learn upon our arrival to our destination in the West Midlands. We hired a driver to take us on the 10-minute journey from Hereford to Ivington. Serendipitously, the driver turned out to be a retired history teacher. He was a storyteller. Perfect for our journey.

As the car drove to Ivington, we immediately discovered that the village is basically an intersection that is put on the map by an elementary school, the house David grew up in, a church, a manor house, a millhouse, and a few small cottages. As a rural location goes, Ivington is both idyllic and historic.

Ivington School was on the main intersection of this storybook village. Some 20–30 students from a several-mile radius would have walked to the school and been schooled within 1 room.[10] The original schoolhouse was built in the nineteenth century, meaning it was from the Victorian era. It has a small stone Gothic rose window at the front. Its simplicity, with six petals, seemed fitting at a school for children, and for us it was unforgettable (Fig. 2.1).

Today the original schoolhouse still exists and remains dedicated to Christian values in a richly rural environment. However, it has been added onto over the years, and the school now serves more students and from a wider radius. At present there are 119 students, many from the nearby market town of Leominster, with over half having special needs such as ADHD. The landscape around the school and residence is very tranquil. The school even has a lovely apple orchard and meadows in its backyard and a small pond.

David's father William Richard David (b.1883–d.*unkn.*) was an educator and the Head Teacher of the school. He was a Welshman.[11] The "industry of service" to which his position was connected

---

[8] David's mother was Florence M. James (b.1881–d.1985). Her grandmother was Florence N. Appleton (b.1860–d.1933).

[9] Note: A correct pronunciation of "Iving" rhymes with "living."

[10] Rees, R.: personal communication to Golbeck, A.L. and Molgaard, C.A. 30 April 2017. Note: At the time this book was written, Rob Rees was a historian.

[11] Note: William Richard David was from Pullandraes, Tondu, Glamorganshire, Wales. His father John (David's paternal grandfather) was a solid blue-collar worker. He was an engine driver. Most likely this meant that he drove

**Fig. 2.1** Ivington School, with David's first residence on the left (2017) [Image credit: Golbeck, A.L.]

was the "City Council." However, the school was jointly sponsored by the Church of England and the education board.

William met David's mother Florence Maude James (b.1881–d.1987) when they both were living in Wolverhampton, which is about 50 miles northeast of Ivington.[12] On the marriage certificate, Florence was listed as a spinster without an occupation. William was 23 years old, and Florence 25, when they married on 8 October 1906 in the parish church of St. George, Wolverhampton.[13] After marrying William, Florence took up the profession of her husband and became an elementary school teacher at the Ivington School under his headship.

The family lived in the residence on the school grounds. It is a brick building detached from, and almost as large as, the main school building itself. The residence would have been provided to the family as part of the parents' compensation package for being the school teachers. It has six rooms, "counting the kitchen, but not counting the scullery, landing, lobby, closet, bathroom, warehouse, office, or shop." The family had a live-in domestic servant.[14]

Four years after the marriage, David was born into this family of educators. A boy had been born right after her parents married, but he died before age 1 and before David was born.[15] This was an era of childhood diseases, and infant and childhood mortality was common. There eventually would be at least two more children, both younger girls. We were unable to find out any more information about David's sisters, except that when David died, one was living in the United Kingdom and the other in Canada.

Also at the intersection known as Ivington, across the street from the school was a church built in the early 1840s and dedicated to St. John [12] (Fig. 2.2).

a railway locomotive, or a steam-powered threshing machine. At the time, these were prestigious occupations in rural communities.

[12] Note: Wolverhampton was David's mother's mother's birthplace.

[13] Note: There were five witnesses to the marriage. Judging from the last names, at least three were from Florence's side of the family. Florence's father Joseph James wasn't there, as he was already deceased. England and Wales Civil Registration Marriage Index 1906.

[14] Census of England 1911.

[15] Utts, J.: personal communication to Golbeck, A.L. 16 September 2021.

**Fig. 2.2** St. John's Ivington where David was thought to be homeschooled by the parson (2017) [Image credit: Golbeck, A.L.]

The church is so small that it is without a spire. Nevertheless, there are two big stories.

One is a mystery story. The St. John churchyard contains the remains of many people from Ivington, but it also has the remains of a princess and prince. The princess, Sophia Mickeladze, was a member of the Russian ruling family deposed in the 1917 revolution. She somehow escaped and was taken in by a family in Ivington. No one knows how she ended up there. Her husband became a prisoner of war and did not make it out of Russia alive. No one knows how his body got to the Ivington churchyard. It is said that the princess had to negotiate directly with Stalin to get his body moved there. The story has long fascinated the children at the Ivington School [13].

The other big story is also a mystery story. It is known that David was homeschooled by a parson. It is likely that he was the parson of St. John's church. David began her schooling at the time of World War I (1914–1918) when she was 5 years old. This was private tutoring, which was a common practice among ambitious parents in rural settings, where the local parson was often the only individual with knowledge of Greek and Latin. These plus mathematics were necessary learning prerequisites for the advancement of a "west country girl" who grew up in the shadow of Offa's Dyke separating England from Wales [14].

Offa's Dyke (or Wall) was a large linear earthwork that followed the border between England and Wales. It lay just a few miles from the hometown of David, trailing across Shropshire and Powys into Herefordshire. It was constructed by a British king of Anglian Mercia (Offa) to keep the Britons of Mercia and the Welsh kingdom of Powys separate from each other and thereby prevent bloodshed. The Wall was 8 feet tall and 65 feet wide and ran from the River Dee estuary in the north all the way to the River Wye in the south.

Raiding along the Wall was incessant, with stiff penalties being exacted for trespass of either of the two tribal groups—a Welshman caught on the English side of the Wall would have his ears cut off, and an Englishman caught on the Welsh side of the Wall would be hung. As a young girl, David would have heard some of the bloodiest stories marking the cultural conflicts of the Welsh and the British, including the role of the Welsh longbowmen in the wars with the French. Hearing stories like this as a child can help make you tough. And David was tough.

**Fig. 2.3** Shelf of books in the chained library in Hereford (2017) [Image credit: Golbeck, A.L.]

The Dyke is still in existence and today is a protected scheduled monument. Supposedly you can walk the entire path in 2 weeks, even if you aren't in great physical shape. Sadly, we didn't have time to try it [15].

The parson taught young David Greek, Latin, and algebra. According to David, he told her at the onset: "Well, you'll of course have to know arithmetic so you better start on algebra. And you can speak English so you'd better start on Greek and Latin" [16]. David later said that, because her father was a classicist, he thought she should be something else, and so she became a mathematician [6] (p. 133).

The marketplace for Ivington is at nearby Leominster (2 miles northeast) or else the larger Hereford (13 miles south). A teacher at one of the local schools was known to explain: "People say Hereford has a draw."[16] One of the attractions is an amazing active Anglican cathedral school and seventeenth-century chained library dating from the era when books were so rare and valuable they were chained to the bookshelf to prevent theft (Fig. 2.3).[17]

David was well known for her love of books. She was a bibliophile. One can imagine David as a young girl being taken to see the books at the chained library and seeing her eyes opened wide.

## 2.2 Head Girl with Prizes

David began her schooling with the local parson at the time World War I was beginning in 1914, when she was 5 years old. By law, she had to begin attending a regular school when she was 10 years old. The law was the 1918 Education Act (The Fisher Act), one of the most important education acts in English and Welsh history. It made school attendance compulsory for all children, both boys and girls, up to age 14. Importantly, it also catalyzed termination of the practice of child labor, which during World War I had been widespread [18]. According to the teacher handbook:

---

[16] Rees, R.: Personal communication to Golbeck, A.L. and Molgaard, C.A. 30 April 2017.

[17] Note: The chained library at Hereford Cathedral is the "largest to survive with all its chains, rods, and locks intact" [17].

> The only uniformity of practice that the Board of Education desire to see in the teaching of Public Elementary Schools is that each teacher shall think for himself, and work out for himself such methods of teaching as may use his powers to the best advantage and be best suited to the particular needs and conditions of the school. Uniformity in details of practice (except in the mere routine of school management) is not desirable even if it were attainable. But freedom implies a corresponding responsibility in its use... However, the teacher need not let the sense of his responsibility depress him or make him afraid to be his natural self in school. Children are instinctively attracted by sincerity and cheerfulness; and the greatest teachers have been thoroughly human in their weaknesses as well as in their strength [19].[18]

Primary education was thus being actively rethought and retooled, with interest in the progressive writings of John Dewey, Edmond Holmes, Maria Montessori, and Susan Isaacs [20].

Around this time, David's parents moved to the town of Beer, which is about 150 miles south of Ivington.[19] Beer is located on Lyme Bay, along the Jurassic Coast, a place that was declared England's first natural World Heritage Site. The young David enjoyed a beautiful locale. Today the town of Beer calls itself "a little piece of heaven" and a "picture-postcard village," with beaches and picturesque white chalk cliffs [21]. David's father became headmaster of the Beer Church Schools, which David began attending [22, 23].

Young David was intelligent and a hard worker. She adjusted well to regular schooling. In September 1920, at age 11, she won a scholarship and started at the selective and highly regarded Colyton Grammar School. This school was established in 1546 (the time of Henry VIII) and is a day and boarding school. It is located in Colyford, Colyton, Devon, which is about 4 miles northeast of Beer.[20] Colyton Grammar School had an excellent headmaster who "sought out those things that needed improvement, and improved them (Fig. 2.4)."

Colyton Grammar School began admitting girls in 1914. This was about 6 years before David entered the school, so David was among the first groups of girls to be schooled there [22–24]. In her last years at the school, David's classes were on the first floor of the building pictured (Fig. 2.5).

There were about 90 students in the school around this time, 40 of which were girls. The building was bursting at the seams, so that in her earlier years at the school, David received instruction in places other than this building [24].[21]

According to a Colyton Grammar School official, today the school is "one of the top state (government funded) Grammar Schools (where pupils sit [for] an entrance exam). The majority of our students attain some of the best grades and go on to attend some of the best Universities in the United Kingdom." Today the student body is comprised of 825 girls and boys ages 11–18 years old.[22]

Fifty years after attending Colyton Grammar School, one of David's classmates recalled that "The standard of education [in David's time] was superb, even under those difficult conditions [having to do with infrastructure], no troubles with discipline, and I for one have never ceased to be thankful for the chance to attend this excellent school and for the riches it added to my life [24, 25]."[23] Also, "Florence David (Beer) and Graham Bond (Axminster) were coming to School on MOTOR-BYKES [sic] at this time! (About equivalent to arriving by helicopter today!)."[24]

---

[18] Note: as quoted in the 1918 Handbook of Suggestions for the Consideration of Teachers.

[19] Note: They lived in Marine House.

[20] David, F.N.: Hand-completed Form of Entrance, Bedford College for Women (University of London), 3 pages, n.d.

[21] Styles, E.: email communication to Golbeck, A.L. 30 July 2018.

[22] Styles, E.: email communication to Golbeck, A.L. 8 August 2018.

[23] Trott, E.M.: handwritten letter to Miss Goate at Colyton Grammar School, 6 pages, 8 March 1976. Note: Edna M. Trott and David were at the same level in 1926.

[24] Note: letter courtesy of Styles, E.

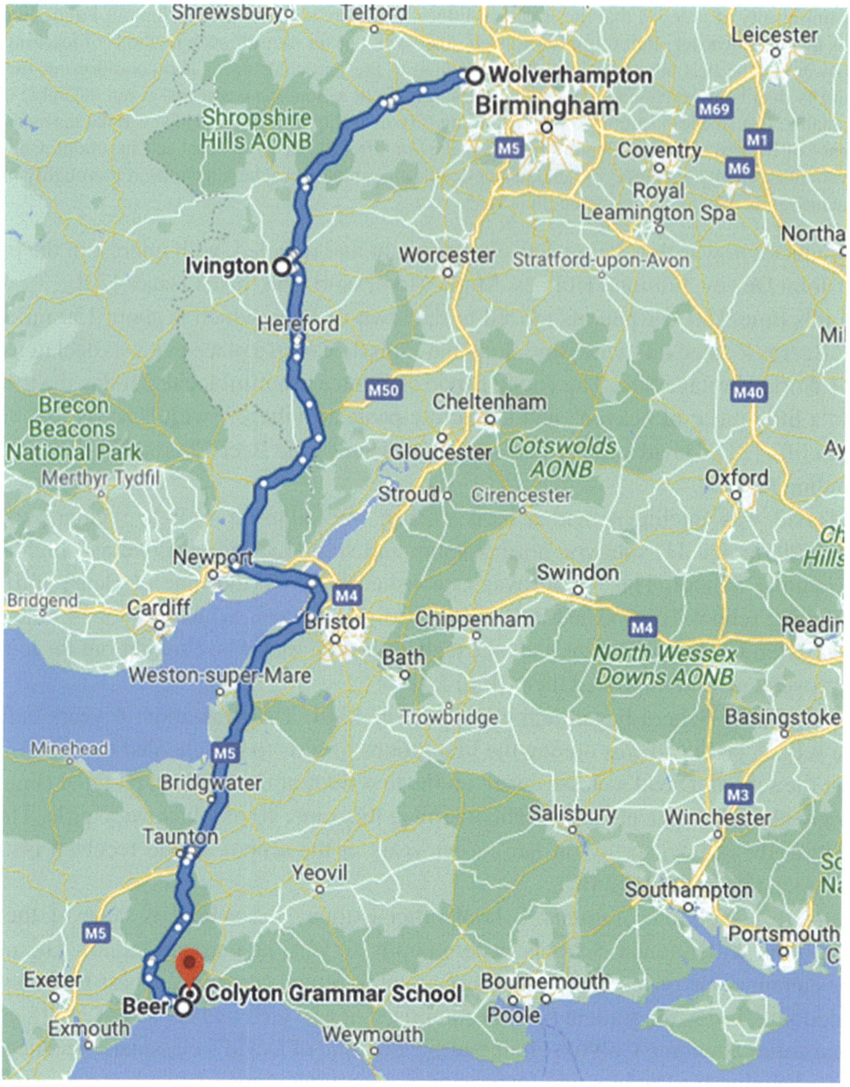


**Fig. 2.4** Google Maps view of David's childhood places [Image credit: Google Maps]

This was just the beginning. David would later take up motorcycle racing. David and W.S. Gosset were friends, and it was Gosset who taught David how to fly fish as a hobby, after she had a motorcycle accident while racing and had to quit [16].

David excelled at Colyton Grammar School and was the Head Girl from 1924 to 1928 [26]. David's peers elected her to the position. Her responsibilities—and also those of the Head Boy—would likely have included representing Colyton Grammar School at events, making public speeches, being a role model, and being a liaison with the school administration and teachers. In the United States, the equivalent would be class president. In England at the time, the position was renewable [27] (Fig. 2.6).

Colyton Grammar School had an Annual Speech Day, where various talks were given and prizes distributed in front of parents and friends of the students. David was a recipient of numerous prizes. At the end of the year in 1926, for example, the headmaster reported that Florence David was one of only four who were successful in the Oxford School Certificate Examination—1st Class Honours. In

**Fig. 2.5** The Colyton Grammar School building where David attended class [Image credit: Styles, E. (Note: At the time this book was written, Dr. Emma Styles was the Funding and Development Officer of Colyton Grammar School Academy Trust.)]

**Fig. 2.6** David as Head Girl of Colyton Grammar School (1927) [Image credit: Styles, E.]

the VIth Form,[25] she was one of the only three who received the Oxford School Certificate Prize—Distinction in Mathematics, the only one who received Distinction in Higher Mathematics, one of the only two to receive Distinction in English, and the only one to receive Distinction in Physics [25]. David and other winners were likely given prizes such as "...green books of the classics with the School Badge and Titles embossed in Gold...", along with other mementos.[26]

On Annual Speech Day in 1930, Colyton Grammar School advanced David as a role model 2 years after she graduated. The dignitary who spoke opined that it was more important to educate a woman than a man. The local newspaper mentioned David by name:

---

[25] Note: In the United States, the VIth Form is equivalent to the last few years of high school.

[26] Trott, E.M.: handwritten letter to Miss Goate at Colyton Grammar School, 6 pages, 8 March 1976.

> [Speaker St. John Ervine, the Irish Playright and Novelist] was a little startled when he heard the Headmaster say that some people asked: "What is the use of educating girls, for they will only get married, and you will have wasted all the money spent on their education?" What a shocking belief that a married woman would be better if she were illiterate—if she were ignorant, and did not know anything at all. It was more important that a woman should be educated than a man, because on a woman depended the first seven or eight years of a child's life. A man, if he saw the children at all, saw them merely as nuisances. If the mother was not intelligent and instructed, how were the rest of us going to have any brains? When he last distributed the prizes at that school a Beer girl—Florence David—secured many of the prizes, and that day a Beer boy—W. Bartlett—took the highest awards (Applause) [28].

In her lifetime, David would defy expectations by not getting married and not becoming a mother. Instead, she would become a renowned professor and highly influential role model for intellectual women. She set the stage for women who came after her to be professors or other kinds of intellectual professionals.

## 2.3 Taking the Bird in Hand

The year 1928 was big for David. She passed the BSc intermediate examination. She was awarded the Devon Major Exhibition (a merit scholarship), which provided 75 GBP a year for 3 years.[27] This was the equivalent to over 4,500 USD in today's currency. She was the first student from Colyton Grammar School to have won this scholarship [23].

It was time to apply for college. David reportedly wanted to go "to London." She was 19 years old. According to David, her mother objected: "My mother was having fits by this time about my going to London... disgrace and iniquity and that sort of thing, so I went to Bedford College for Women" [16]. David may have been referring to UCL when she said "London," because Bedford College was also located in London. Founded in 1849, Bedford College was the oldest women's college in the United Kingdom [29].

In fact, what David really wanted to do was to attend Newnham College at the University of Cambridge. Newnham is "the lively and sociable women's College at the heart of the University of Cambridge. Founded in 1871, [it is] the oldest college run by women, for women" [30]. However, David was turned down by Newnham, which is how she decided to attend Bedford College.[28]

Here is the story. On 9 April 1928, David's headmaster at the Colyton Grammar School wrote the following letter to the Principal of Bedford College for Women:

> One of my girls, F.N. David, has recently taken the entrance examination at Newnham [College - University of Cambridge] without success: Could you take her at Bedford College next October? She has passed the - - - Inter. B.Sc. in Pure & Applied Mathematics Physics & Chemistry & is a girl well above the average in every way. Captain of the [field] hockey XI & Senior Girl. She has proved herself an excellent leader & being a girl of strong personality. I am confident she would be a credit to the College. If you accepted her she would, with your approval, like to try again for Newnham next spring hoping to get a scholarship. Otherwise she could stay on at Bedford College & take her honours degree. I hope you will be able to take her as she reaches the age of 19 in August & I should find it difficult to keep her in our school for another year.[29]

The Bedford College registrar's office replied that most of the vacancies for the upcoming session had already been assigned, but there were a few unassigned seats set aside for nonresident students; perhaps the principal would be willing to consider an application from David for such a seat. The assistant registrar invited David to fill out an application and pay the registration fee of 1 guinea.[30]

---

[27] Valentine, A.: email communication to Golbeck, A.L. 11 May 2017. Note: At the time this book was written, Annabel Valentine was Royal Holloway University Archivist.

[28] Valentine, A.: email communication to Golbeck, A.L. 19 May 2017.

[29] McKay-Ohm, D.: handwritten letter to Bedford College principal, 2 pages, 9 April 28.

David followed up the next day with her application. She explained to the registrar:

> I am desirous of obtaining admission for the session 1928-1929, during which period I shall try to obtain admission to Newnham College, Cambridge. If, however, I should be unsuccessful, with the Council's permission, I shall wish to continue my studies at Bedford with a view to taking an honours degree either in Mathematics or Science.
> I should be very grateful if you could grant me admission as a non-resident student in October.

We note that David was still signing her name "Florence N. David."[31]

David included her "Form of Entrance," i.e., application for admission, with her letter. She asked for entrance to the winter term. She listed the following on the Examination Record line:

> Exemption from London Matric. on Oxford School Certificate.[32]
> 1st Class Honors.
> Distinctions in Mathematics and Higher Mathematics.
> Credits in English, History, French, Chemistry, Physics July 1926.
> London External Inter. B.Sc. in Mathematics, Applied Maths, Physics, Chemistry Nov.-Dec. 1927.

David indicated she did not want to "come into Residence." She would be applying for admission to another college, but "not until 1929." She proposed to follow "the first year of a course leading up to an Honours degree in Mathematics." Signing her application were her father, grammar school headmaster, and the Vicar of Beer.[33]

In his letter of support, David's headmaster spoke to more than David's academic preparation. He wrote: "I consider she is suited in every way to enter the college, and I feel she will have an excellent influence on many with whom she comes in contact."[34]

On the one hand, David and her headmaster were honest. On the other hand, Bedford College didn't take it well that David wanted to only spend a year at Bedford and then go on to Newnham. Here is what the principal of Bedford told David:

> Although you have passed the External Intermediate Science examination and are therefore qualified for admission to the College as a non-resident student without further examination, I cannot hold out very much hope of being able to accept you as a student here, since you propose to re-enter for the Newnham Scholarship in 1929.
> We have a very long waiting list of candidates who wish to read for degrees of this University and who qualified for admission on the Entrance Examination held in March. It is not fair to them to accept late applicants who propose to use the College as a stepping stone to another university.
> I will let you have a definite answer early in June but, in view of the uncertainty and of the fact that, if you were admitted, your instruction would have to be limited to the First Year Honours Course in Mathematics (no special tuition for the Scholarship Examination could be arranged here), you may prefer to make other arrangements. I shall be glad to know whether you decide to withdraw your application or will like the question considered in June.

The principal was clearly insulted.[35]

What was David to do? There seemed to be only one reasonable choice. David decided to take the bird in hand:

---

[30] Bedford College Assistant Registrar: letter to McKay-Ohm, D. 10 April 1928.

[31] David, F.N.: handwritten letter to Bedford College principal, 2 pages, 18 April 1928.

[32] Note: The School Certificate examination was required by national law between 1918 and 1951. Taken at age 16, it qualified students to leave school. The matriculation exemption was granted when the student obtained credits in five subjects that included mathematics and English. See: 31 October 2013 A* to G grades will be abolished in revamp of GCSEs: Exams will given [sic] numbers from one to nine while papers based on pupils' ability will also be axed [31].

[33] David, F.N.: Hand-completed Form of Entrance, Bedford College for Women (University of London), 3 pages, n.d.

[34] McKay-Ohm, D.: handwritten letter to Bedford College, 1 page, 17 April 1928.

[35] Bedford College Principal: letter to David, F.N. 1 May 1928.

> I am sorry you think it will not be possible for me to come for a year, but I quite understand that I should be leaving a vacancy in your 1st Year Honours Course if I was to leave at the end of the year 1929.
> In view of this fact if you would be so kind as to give me entrance in October 1928, I should like to finish out my course at the college and take my finals in 1930 or 1931, whichever you think best.
> I have a fairly good scholastic record and have been girls captain of my school for several years. I shall be very grateful if you can find it possible to give me entrance in October 1928.[36]

David was strongly supported with a letter from her headmaster to the principal. He would have known David's abilities, because he was the math master in addition to being the headmaster. [37] The headmaster repeated some of David's academic accomplishments and David's request.

At the same time, the headmaster continued to make it clear that he would have preferred David to go to Newnham:

> ...I was anxious for her to go to Newnham, if it had been possible, but we are perfectly willing to fall in with your conditions and for her to remain at Bedford College. Do you think, in these circumstances, you will be able to offer acceptance? Otherwise I must look elsewhere and do so speedily...Would it have been in order for Miss F.N. David to have joined your Inter. B.Sc. class in September and have taken the Newnham entrance in the Spring of 1929? Or is it possible for her in any other way to have tried for Newnham and to stay [unclear] if she had failed? I want [unclear] of course for her to be accepted at Bedford College.[38]

In spite of the headmaster's continuing references to Newnham, the bird-in-hand approach worked.

Bedford College offered David a nonresident student position to start in October, under the condition that she "definitely decide to read for the B.Sc. Special examination." David would be expected to take Special (or Honours) courses and spend 3 years preparing for the final exam.[39]

At the same time, the Bedford College principal remarked on the "very long waiting list of candidates who qualified for admission ...only in the event of there being places available late in July that a candidate wishing to use this College as a stepping-stone to another University could be considered." The principal offered "only very slight hopes" that they could accept David for only 1 year.[40]

David decided to not use Bedford as a stepping stone and committed to remain at Bedford. She received her acceptance the next day. She was grateful.[41] The only thing left to do was to apply to go into residence when residence became available, meet with the principal, and fill in a form letting the college know where she would immediately be living.[42] The only other thing left to do was the rest of her productive career.

## References

1. Chretien, J.: Order worthy? National Post 15 July (2009) https://www.pressreader.com/canada/national-post-latest-edition/20090715/281917359074501 Accessed 22 August 2018
2. Jackson, M.D.: The Order of Merit 1902–2002: One hundred years of matchless honour by Stanley Martin, CD. Canadian Monarchist News (Summer) 26, 15 (2007)
3. Rayment, L.: List of members of the Order of Merit (2017) http://www.leighrayment.com/orders/orderofmerit.htm Accessed 22 August 2017

[36] David, F.N.: letter to Bedford College principal, 3 May 1928.

[37] Trott, E.M.: handwritten letter to Miss Goate at Colyton Grammar School, 6 pages, 8 March 1976.

[38] McKay-Ohm, D.: handwritten letter to Bedford College principal, 4 pages, 4 May 1928.

[39] Bedford College Registrar: letter to David, 8 May 1928.

[40] Bedford College Principal: letter to headmaster, 8 May 1928.

[41] David, F.N.: handwritten letter to Bedford College principal, 2 pages, 9 May 1928.

[42] Bedford College Registrar: letter to David, F.N. 18 September 1928.

4. American Statistical Association: Florence Nightingale: The lady with the data (2018) This is Statistics http://thisisstatistics.org/florence-nightingale-the-lady-with-the-data/ Accessed 12 September 2018
5. Hassmiller, S.: Parting thoughts: 10 lessons learned from Florence Nightingale's life. American Journal of Nursing. Off the Charts (2017) https://ajnoffthecharts.com/parting-thoughts-10-lessons-learned-from-florence-nightingales-life/ Accessed 12 May 2017
6. Reid, C.: Neyman - From Life. Springer-Verlag, New York (1982)
7. O'Connor, J.J. and Robertson, E.F.: Florence Nightingale David. School of Mathematics and Statistics, University of St Andrews, Scotland (2004) https://mathshistory.st-andrews.ac.uk/Biographies/David/ Accessed 13 March 2023
8. Salsburg, D.: The Lady Tasting Tea: How Statistics Revolutionized Science in the Twentieth Century. Henry Holt and Company, New York (2001)
9. Leonard, T.: The life of a Bayesian boy (2017) http://thomashoskynsleonardblog.blogspot.com/2021/04/the-life-of-bayesian-boy-tom-leonards.html Accessed 4 July 2023
10. Box, G.E.P.: An Accidental Statistician: The Life and Memories of George E.P. Box. Wiley, New York (2013)
11. Mills, A.D.: Dictionary of English Place-Names. Oxford University Press, Oxford, United Kingdom (1998)
12. Herefordshire Council: Monument detail, Church of St John, Ivington, Leominster. Herefordshire through time (2018) https://htt.herefordshire.gov.uk/her-search/monuments-search/search/Monument?ID=18236 Accessed 9 September 2018
13. Hereford Times: Ivington mystery of Royal couple. 21 February (2005) https://www.herefordtimes.com/news/5649284.ivington-mystery-of-royal-couple/ Accessed 13 March 2023
14. History Extra: A brief history of Offa's Dyke. BBC History Magazine and BBC History Revealed (2023) https://www.historyextra.com/period/anglo-saxon/a-brief-history-of-offas-dyke/ Accessed 13 March 2023
15. Dunn, M.: Offa's Dyke map booklet. Cicerone (2021) Accessed 9 March 2021
16. Laird, N.M.: A conversation with F.N. David. Statistical Science 4(3), 235-246 (1989)
17. Hereford Cathedral: Chained Library (2023) https://www.herefordcathedral.org/chained-library Accessed 13 March 2023
18. McCulloch, G.: The Great War and education. The Telegraph 4 August (2014) https://www.telegraph.co.uk/education/educationopinion/11005281/The-Great-War-and-education.html Accessed 9 December 2018
19. Great Britain Ministry of Education: Primary Education: Suggestions for the consideration of teachers and others concerned with the work of Primary Schools. HMSO, London 9 (1959)
20. Gillard, D.: Education in England: a history. May (2018) http://www.educationengland.org.uk/history/chapter04.html Accessed 13 March 2023
21. VisitBeer: A little piece of heaven (2023) http://www.beer-devon.co.uk/ Accessed 14 March 2023
22. Pulman's Weekly News: Miss F.N. David's success. 9 October (1935)
23. Pulman's Weekly News: Beer student's distinction: Doctorate in philosophy 19 July (1938)
24. Gosling, G.: Colyton Grammar School. Old Colytonians Association (1978)
25. Pulman's Weekly News: Colyton Grammar School, Speech day and prize distribution 2 December (1926)
26. Pulman's Weekly News: Beer student's distinction: Doctorate in philosophy. 19 July (1938)
27. Wikipedia: Head girl and head boy (2018) https://en.wikipedia.org/wiki/Head_girl_and_head_boy Accessed 22 May 2018
28. Pulman's Weekly News: Women's education: More important than men's. Mr. St. John Ervine's address at Colyton School Speech Day December (1930)
29. Ronson, J.: Pioneering women's education at Bedford College. Archives Hub, September 30 (2019) https://blog.archiveshub.jisc.ac.uk/2019/09/30/pioneering-womens-education-at-bedford-college/ Accessed 17 August 2023
30. University of Cambridge: Newnham College (2017) http://www.newn.cam.ac.uk/ Accessed 19 May 2017
31. Levy, Andrew: A* to G grades will be abolished in revamp of GCSEs, Daily Mail 31 October (2013) http://www.dailymail.co.uk/news/article-2482534/GCSE-A-G-grades-abolished-exam-revamp.html Accessed 8 June 2018

# 3 Learning (1928–1931)

*My only quarrel with [David] is that she overworks but she replies she enjoys it.*

Bedford College Instructor: report card for David, F.N. (1928)

## 3.1 Zeroing in on Mathematics

David headed off to Bedford College for Women. It was a 160-mile journey from Beer to London.

We can easily imagine David making the trip from home to college on her motor—byke. These bikes did a lot for advancing the rights of women. They helped to promote for women the seeking of adventure, the freedom of the road, the wearing of loose clothing, and a progressive public image. Susan B. Anthony said at the end of the nineteenth century that "The bicycle has done more to emancipate women than any one thing in the world" [1]. There is a fun site on the web that calls women who ride motorcycles "badass women who were not afraid to take the rides of their lives" [2]. No matter whether she went on her motor—byke or took some other form of transportation, in the year 1928, David was taking the ride of her life to London.

Bedford College was a serious place for women. It started as the nonsectarian Ladies' College for liberal education in 1849.[1] The college had some influential connections. For example, Erasmus Darwin, Charles' older brother, served on the original Council, chairing it for 18 years and holding the college's original funding in trust. The vision was for the college to be a place for stimulating the best use of women's mental faculties, requiring them to take "masculine" subjects like Latin and mathematics, and not the "lighter subjects only."

In the United Kingdom, Bedford was the first institution of higher education for women, attracting many Unitarians and feminists. It had women on its board of governors from the beginning. It was a catalyst for the equality of women in higher education in particular and society in general. In 1859, the name was changed from the Ladies' College to Bedford College. There were many "firsts" at the college: first woman professor at an English university (Caroline Spurgeon); first Social Sciences department in the United Kingdom, established in 1918; first woman vice-chancellor; first women factory inspectors under the 1913 Health Act; first woman to be appointed to the Companionship of the Order of the Bath; one of the first two women Fellows of the Royal Society; fourth woman

[1] Note: The college in London started at the address 47 (now 48) Bedford Square. In 1974 it moved to larger premises at 8 and 9 York Place (off of Baker Street). In 1908 it moved to Regent's Park.



A. L. Golbeck, *Florence Nightingale David*,
https://doi.org/10.1007/978-3-031-77768-4_3

chairperson of the Trades Union Congress (Marie Patterson); and first art school in England where women could draw from life. Also, Bedford launched the fashion for bobbed hair. When the first bobbed head was seen—at Bedford College—it is recorded that it nearly caused "other students to faint."

In 1900, Bedford College became part of the University of London, which began granting degrees to women in 1878. When David started at Bedford, there were about 640 students, all women.[2] David went to the college when it was in the lovely Regent's Park location. Regent's Park is one of London's Royal Parks. It has a lake and many beautiful gardens and playgrounds [3].[3] Bedford was able to secure this prime location, accessed through York Gate, by an unexpected gift that allowed the college to lease land from the Crown. David was there at a more progressive time than when the College started. Society was understanding the advantages of higher education for women, and there was a fundraising push. The college commissioned British architect Basil Champneys (b.1842–d.1935) to design the buildings. He had already designed many notable university buildings, chapels, and museums in the United Kingdom (including at Cambridge and Oxford). "There were seven blocks, including two for science, in an L-shape that had the residential halls on the shorter side" [4–7].

David worked hard at Bedford College from 1928 to 1931. She could have studied journalism, English language and medieval literature, English literature, Dutch studies, French, Italian, German, modern and medieval history, ancient history, Greek, Latin, mental and moral science, psychology, mathematics, geography, geology, organic chemistry, inorganic and physical chemistry, physics, botany, zoology, biochemistry, physiology, nursing studies, social studies, sociology, or economics [4].

With all of these subjects to choose from, David chose to specialize in the "masculine" subject of pure mathematics.[4] Looking back many years later, she said she didn't do anything but mathematics while at Bedford, and she "didn't like that very much" [8]. David loved mathematics, but she had broader interests.

David was, in fact, one of a growing number of young women in England at the time who were seeking training in the sciences. The times were changing, and women were overcoming the former "outrage in the press including doctors saying that women wouldn't be able to cope with the [science] subjects and they would become less feminine and may not be able to have children if they studied them."[5] When we visited the Bedford College archive, it was held at Royal Holloway University. We found a set of posters—proudly on display—that featured women in the sciences at Bedford (Fig. 3.1).

As a student, David received a report card three times each year. These times corresponded with important religious observances: Michaelmas, Lent, and Easter. Grading was done with Greek rather than English letters. On her report from 1928 to 1929, she received an $\alpha$- ("Improving very rapidly. My only quarrel with her is that she overworks but she replies she enjoys it"), an $\alpha$ ("Does excellent work"), and another $\alpha$ ("A very good student, keen on her work").[6] On her report from 1929 to 1930, her grades dropped a bit due to temporary health issues: She received a $\beta$++ ("A succession of bad colds has been against her –"), a $\beta$+, and a $\beta$.[7]

---

[2] Valentine, A.: email communication to Golbeck, A.L. 17 May 2017.

[3] Note: In 1965, the college became coeducational. In 1985, it merged with Royal Holloway College—another college that began 40 years later as women's college but at the time this book was written was coeducational and located in Egham, Surrey; the buildings used by Bedford College in Regent's Park were being used by the University of Illinois' Regent's University London.

[4] Valentine, A.: email communication to Golbeck, A.L. 11 May 2017.

[5] Royal Holloway University: Display, Women in Science at Bedford and Royal Holloway Colleges, Royal Holloway University Archive Room.

[6] David, F.N.: report form, Bedford College, 1928–1929.

[7] David, F.N.: report form, Bedford College, 1929–1930.

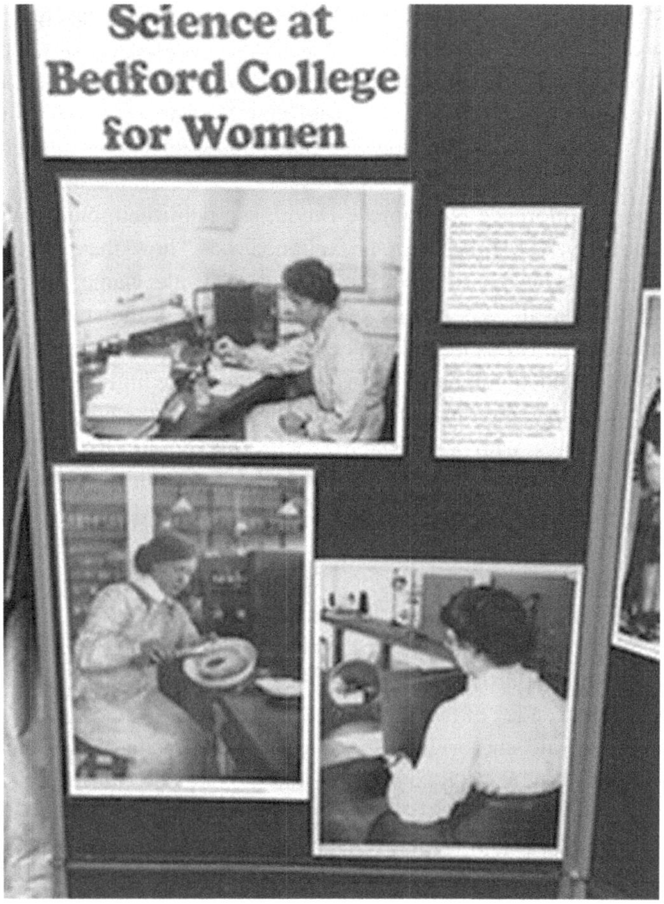


**Fig. 3.1** Display indicating pride in "Science at Bedford College for Women" (2018) [Image credit: Golbeck, A.L.]

In spite of her love for mathematics, David later said she didn't like Bedford College very much: "I didn't like the people and I suppose I was a rebel in those days but I don't look back on it fondly…" [8]. What David did like during her college years was going to the London theater every night, such as to the Old Vic, where because she was a student it only cost her 6 pence to see a show ("about 3 new pennies" which would be about 50 cents in 2017 USD). David wasn't alone in her love of the theater: As early as 1915, one of the most popular outside entertainments for students at Bedford was to attend a matinee in the theater district [4].

In June 1931 at age 22, David completed her education at Bedford College by obtaining a pass in her full BSc Special Degree Examination in Pure Mathematics. Later in life, she would serve as a governor of the college. The main function of the governors was to participate in the autumn annual general meeting, which is when the year's activities were reported, but the governors were also welcome at social functions.[8] The governors were the keepers of the constitution of the college, gave advice, and oversaw the college's management council [4].

David's assignment as a governor didn't improve her opinion of her time at the college [8]. She started as a governor in 1958 and continued through at least 1968. In fact, she may have continued on

[8] Bedford College Secretary: letter to David, F.N. 25 June 1956.

until her death in 1993, as there wasn't a set term for the position and at least one person is known to have left the role by passing away. It wasn't clear that David attended any of the meetings—at least, she isn't listed as attending [4].[9] David would later be mentored by the statistician K Pearson who, when once asked how he had so much time to write, said "...I never answer a telephone or attend a committee meeting" [9]. Perhaps K Pearson's attitudes about telephones and committee meetings rubbed off on David.

Although her name was Florence Nightingale David, she published under the name F.N. David. Her friends called her David, which Neyman explained was "how they address you in English universities" [10] (p. 132). One possible reason for not using the name Florence is that a name referring to Florence Nightingale was simply overwhelming given the stature of *the lady with the lamp*. Another possible reason is pushback to the unsymmetric practice at Colyton Grammar School where the lists of honors included first names of women but only the initials of the first names of men. A third possible reason is carryover of the Bedford College practice where the women were almost always called by their surnames, with or without the "Miss" [4] (p. 41). UCB colleague Eric Lehmann (b.1917–d.2009) reported that David detested her first and middle names [11] (pp. 116–119).

In any case, David's formative years did not occur at Bedford College. These lay ahead at UCL at the hands of Karl and Egon Pearson and many others in the great British tradition of numeracy, statistics, and biostatistics such as Jerzy Neyman (b.1894–d.1981), William Sealy Gosset (b.1876–d.1937), R.A. Fisher (b.1890–d.1962), Norman Johnson (b.1917–d.2004), Maurice Kendall (b.1907–d.1983), D.E. Barton (b.1926–d.*unkn.*), Colin Mallows (b.1930–d.2023), J.B.S. Haldane (b.1892–d.1964), and Maurice Bartlett (b.1910–d.2002). David had many "special relationships" in the quantitative fields on both sides of the Atlantic.

Later in her life, David said she originally wanted to be an actuary. She said it was for some "obscure reason" [8]. Earlier she had said it was to avoid the basically-female role of teacher [10] (p. 133).

It is actually no wonder that David, who loved mathematics and problem-solving, wanted to be an actuary. It is a field that is focused on applied probabilities. "Actuaries are professionals who apply mathematics to financial problems. They evaluate the financial implications of contingent events, in other words, events that are not certain to occur" [12]. They are people "trained in evaluating the current financial implications of future contingent events. It is the actuary's job to assist in the scientific analysis and quantification of risk...[they] build models which make best use of the available information. In general, actuaries excel in problem solving" [13]. Actuarial science would have been a perfect field for David.

The context in the United Kingdom when David was growing up might further explain her interest in becoming an actuary. Health insurance and unemployment insurance were big news. National Health Insurance was introduced by legislation in 1911 ("An Act to provide for insurance against Loss of Health and for the Prevention and Cure of Sickness and for insurance against Unemployment, and for purposes incidental thereto"). By 1936, it covered 19 million wage earners and their families [14]. The Unemployment Insurance Act was introduced by legislation in 1920, providing payments over 15 weeks. It was passed at a time when there was very little unemployment. However, the next year, there was very high unemployment with huge increases in costs to the Treasury [15, 16].

David was in a structured school at this point. It could be conjectured that these big news developments in insurance were discussed in her classes and at home. By the time David moved to London to start at Bedford College in 1928, there were over 2 million unemployed in Britain. There would have been considerable use for actuarial expertise. In fact, it has been written that

[9] Valentine, A.: email communication to Golbeck, A.L. 17 May 2017.

"Unemployment was the dominant issue of British society during the interwar years," evidenced by the fact that there were five unemployment insurance acts passed between 1920 and 1930 [17].

We can still remember from when we were young that there were gender-specific job advertisements in the newspapers: Help Wanted Male and Help Wanted Female. Actuary jobs would have fallen under Help Wanted Male. David recalled that, at the time she was interested in the profession, "the actuaries would only take men." David reportedly applied under the name F.N. David to be an actuary and was offered the job, but when the employer found out she was a woman, they withdrew the offer [10] (p. 133.). She was told that she couldn't have an actuary job because there were no women's restrooms [18].

In the United Kingdom in the early 1900s, actuarial practice had expanded in scope and complexity, and the number of actuaries was increasing [19]. However, it was only in 1923 that the first woman received her qualifications as an actuary, and by the early 1950s, there were only nine women who had passed the extra exams and qualified as Fellows of the Institute of Actuaries [20]. Women were not breaking into this profession at the speed that they were breaking into other professions.

## 3.2 K Pearson and Women

K Pearson was the professor and head of the statistics department at UCL. He already had an immense reputation by the time David met him in 1931. One eminent sociologist referred to the applied statistics department at UCL as "the home of the gods" and wrote that K Pearson was a "hero of Asgard" and was even "Thor himself" [9].

David must have known that K Pearson was a truly great mathematical statistician. Today he is remembered for being one of the founders of modern statistics. Much of his work is foundational to what is studied in first statistics courses. He was instrumental in developing the theory of linear regression and correlation, including the Pearson product moment correlation coefficient which was named after him. When students study regression, they often hear about regressing sons′ heights upon their fathers′ heights. This example came straight from K Pearson's empirical work. He is also known for his work on methods of moments and classification of continuous probability distributions. Particularly important was his work on the exponential family of distributions, as it underlies the theory of generalized linear models [21]. K Pearson is also credited with developing the ideas of chi distance, p-value, foundations of statistical hypothesis testing theory, statistical decision theory's chi-squared test, and principal components analysis [22–25].

K Pearson had a long history of interest in gender issues and had promoted several women in mathematics by the time David met him. His interest in women in science developed early in his life. As a 28-year-old, K Pearson co-founded the radical Men and Women's Club. This was comprised of equal numbers of men and women who met over a 5-year period (1885–89) to promote debate about gender norms. According to one author, K Pearson

> ...believed that women were the key to national progress. In the club's inaugural paper, "The Woman's Question," he reflected on what change would occur should women gain access to education, professional and political representation...Pearson went on to become the premier voice on the "Woman Question" during the 1890s. His writings were read in Britain and America, and feminists on both sides of the Atlantic viewed him with much respect, using much of his rhetoric to push for legislative reform for women [26].

Members of the Men and Women's Club wanted to establish answers to the woman question on the basis of objective evidence, not just their own opinions. K Pearson pursued detailed statistical analyses in his inquiries about the sexes [27].

Over the years, K Pearson had a number of women working for him in his biometrics laboratory [28]. There were 5 women out of 14 employees in 1908. In fact, it has been said that K Pearson was a leader in offering scientific work to women who wanted to do "head work" instead of the manual labor positions typically held by women at the time. It has also been said that he was an example of The New Man who, with generosity and vision, was moving The New Woman forward. At the same time, K Pearson acknowledged there were economic factors that contributed to his hiring women. They were cheaper than men, and even cheaper than women working as assistant mistresses in secondary education schools [27, 29].

At least two of the women were, like David, graduates of Bedford College. They were Alice Lee (b.1859–d.1939) and Ethel M. Elderton (b.1878–d.1954).

Three years after the closing of the Men and Women's Club, K Pearson met a bright and mathematically gifted woman named Alice Lee. Lee had attended the first class in higher mathematics at Bedford College, was the first student to receive a BSc there (1884), subsequently earned a BA from Bedford (1885), and then was one of the first women to earn a PhD from UCL (1901). She stayed on at Bedford College until 1916, first as an assistant teacher in mathematics and physics and then rising to the position of research lecturer in applied mathematics [27, 29, 30].

The meeting between K Pearson and Lee happened sometime after he wrote an article in *The Pall Mall Gazette* attacking the academic standards at Bedford College and arguing that the school should be closed. His basic objection was to one person teaching two very different subjects, e.g., geology and botany, which was happening at Bedford due to shortages of resources. In fact, Lee herself was involved in teaching Greek and Latin in addition to mathematics and physics. Lee vigorously defended the college. Something about Lee must have impressed K Pearson, and vice versa, because after this heated exchange she went to study and work under him in his biometrics laboratory.

Lee became known for her work on skull capacity, especially her empirical rebuttal of the implicitly sexist theory that larger skulls (generally associated with men) were associated with greater intelligence or eminence. She was also known for her body of published work with K Pearson on the mathematical contributions to the theory of evolution. She eventually retired from K Pearson's laboratory in 1927, 4 years before David arrived on the scene [30]. Lee had lived on low women's wages and didn't get into the university pension fund because it started too late for her to join. K Pearson successfully advocated for her to receive a civil list pension, in recognition of the "first class research" she had done in her career [27].

The other woman graduate of Bedford College who worked for K Pearson was Ethel Elderton. Lee had been Elderton's mathematics instructor at Bedford College. Lee recommended and introduced Elderton to Galton in 1905, and Elderton became K Pearson's statistical assistant, working her way up to assistant professor, Galton scholar, and fellow.

Elderton was known for her controversial paper arguing that predispositions to alcoholism were generally inherited [31]. She is little known for her important observation that induced abortion was a major contributor to the decline in birthrate in England [27]. Elderton was working for K Pearson when David appeared on his doorstep looking for an opportunity. Elderton was an assistant professor, Galton research fellow, fellow of the college, and reader in eugenics.[10] When K Pearson retired in 1933, she stayed on as a reader in applied statistics, but she ended up retiring 2 years later at the end of 1934–1935. K Pearson's son Egon, who had been a senior lecturer, was given the department head position upon his father's retirement, even though Reader Elderton had a higher-level position. David would have known Elderton for 2 years at UCL, as they worked together to support K Pearson's vision of a "quantitative science of man."[11]

[10] Note: Elderton was promoted to reader in 1932.

[11] Mawer, A.: letter to Pearson, E. 24 April 1933. Note: Mawer was University College London provost.

Lee, followed by Elderton, and later followed by David, wanted to do brain work. Brilliant women such as these believed that their innate abilities were equivalent to those of men:

> ... to this end they were concerned with the education of women and their participation in new fields of labor, and the extension of the franchise. Clearly the waste of human potential entailed in denying basic freedom to women was against the best interests of the race, and must be resisted. The future of the race lay with women, too, assisting in the evolutionary process, even if the need to fit women to their present environment for the sake of the future involved the indirect choice of a career over a direct response to the eugenic creed in marriage and child-bearing. Whatever its particular attractions to the individual, belief in eugenics clearly gave the women workers of the laboratory something which remained with them after the novelty of their new occupation had worn off, when they realized that for the rest of their working life nearly all their time must be given to it, for wages insufficient to support them independently.

In a world where science and higher education were dominated by men, these women struggled and fought for educational opportunities. K Pearson provided an environment that allowed them to participate in work that they found meaningful. This work promised to provide scientific solutions to societal problems, including the problems women faced in society [27].

David may have known about K Pearson's actuarial science interests through her Bedford College connections. As previously mentioned, David recalled many years later that someone had told her about K Pearson, that he had done some actuarial work, and that was why she crashed her way in to see him at UCL. For David, that someone may have been an individual or individuals at Bedford College, as both Lee and Elderton were Bedford College alumni and K Pearson employees, and it may even have been Elderton herself.

It was surely known at Bedford College that K Pearson hired women to do statistical work and that he had well-developed and long-held interests in women's equality. Although David claimed it was luck, or possibly that she went there by fate, she may have had just enough information about K Pearson to assess the odds being in her favor, that he would be receptive to her when she unexpectedly appeared to him out of nowhere, and then that the two would have shared intellectual interests.

David was correct that K Pearson had done some actuarial work. It was within the university. In his earliest teaching, he "discussed graphical statistics and the mathematical theory of probability with a particular interest in actuarial methods." Later, when he took over for Galton in the eugenics laboratory at UCL, he developed special methodologies that were "underpinned by the use of actuarial death rates" [32]. K Pearson was also a member of the Actuaries' Club [33]. This is the oldest actuarial club for dining. This by-invitation-only club predates the Institute of Actuaries [34].

Also, Elderton's brother William Palin Elderton (b.1877–d.1962) knew K Pearson and was an accomplished actuary. In fact, the brother met K Pearson while an actuarial student, computed the first chi-square tables (1902), and published an actuarial monograph on the Pearson curves (1907). The brother was president of the British Institute of Actuaries (1932–34) and published in the statistical methods and theory journal *Biometrika* that K Pearson co-founded in 1901 with zoologist Raphael Weldon and Francis Galton.[12] Also, the brother became chairperson of the Biometrika Trust (1935) [35, 36].

## 3.3 Crashing Her Way In

David knew she could take and pass the actuarial exams to navigate around some of the gender issues that seemed to be blocking her entry as a woman to the profession. But her path took another turn.

[12] Note: They founded the journal because of the great difficulties they were having publishing in other journals. Galton's role was essentially a patron.

She decided to check out UCL and ended up going there to study and work.[13] David E. ("Dick") Barton put it this way: "Finding [the actuarial] profession effectively closed to women (as it was then) she did the next best thing and joined K Pearson at UCL as a research assistant [37, 38]."[14]

At the time, K Pearson was professor and head of the department of applied statistics and eugenics at UCL. He was also an honorary fellow of King's College, Cambridge.[15] Founded in 1826, over 100 years before David arrived on its doorstep, UCL was the first university in London and the first secular university in England (Oxford and Cambridge were religious-based universities). Its origins were from the liberal experience and schools of Scotland.

UCL was established by individuals inspired by English philosopher and political radical Jeremy Bentham (b.1748–d.1832), known for consequentialist thinking (e.g., actions should be evaluated on the basis of their consequences in producing happiness among those affected) and legal positivism (e.g., there are human rights that exist independent of government) [39]. Bentham was age 78 when UCL was founded as an English alternative to "Oxbridge" (Oxford and Cambridge) educational practices and theories.

Bentham's ideas were that education should be made generally available, not just to the wealthy and not just to those of a certain religion. David said: "[UCL] was founded for 'Turks, Infidels, and such as do not profess the 39 Articles,' because in those days to go to a university you had to say that you believed in the 39 Articles of the Church of England. So University College was founded, and a year after that King's College in London was founded as a Christian antidote. Women were admitted to lectures very early on at University College. So they theoretically had no impediment bar the usual male one" [8].

In some ways Bentham was the spiritual founder of UCL (Fig. 3.2).

"Auto-icon" is a little-known word in American English. The Oxford Living Dictionaries says that an auto-icon is "A dead body which is preserved, clothed, and displayed as though still living, as

**Fig. 3.2** Golbeck, A.L. with auto-icon Jeremy Bentham (2017) [Image credit: Molgaard, C.A.]

[13] David, F.N.: First Entry Form for University of London, University College, 1931–1932; Re-Entry Form 1932–1933. Note: At this time, David was living at 1A Callow St. in Chelsea.

[14] Note: According to Mallows, Barton died at a young age.

[15] University College London Calendar 1931–1932.

a memorial to the deceased." The dictionary tells us that the origin of the word was associated with Bentham [40]. This is because Bentham willed that his body be dissected in public and then preserved as an auto-icon.

Thus, one can imagine that every day when David went to work at UCL, she was in the presence of remains of Bentham, sitting in a handsome armoire in a foyer of the South Cloisters which is part of the UCL Main Building. Bentham's skeleton was dressed up in his clothing stuffed with hay, with a wax head that included some of Bentham's own hair. Because the actual head wasn't well preserved and came out looking very distasteful, it wasn't displayed but stored in a locked cabinet out of view.

In fact, the armoire inscribed with the name Jeremy Bentham is still opened each workday in order for the public to view the Bentham auto-icon. We found it to be a bit creepy to walk through the South Cloisters and encounter it. At the same time, the auto-icon is an effective reminder of the man and his spiritual journey, and it makes you want to learn more about him and his philosophies. In 2018, the auto-icon was on display at the Metropolitan Museum in New York presumably to extend its audience [41].

The Foreword to a leading book on UCL describes the university in this way:

> Institutional histories are rarely riveting.
> But UCL is, in this as in so many other areas, a clear exception. It has throughout its time been a university of innovations and firsts, with no fewer than 9 Nobel Prize winners,[16] [42] to show for it. And from the beginning its focus has been international: indeed that very word was coined by Jeremy Bentham, who is still today a dominating presence at UCL (not only intellectually but physically).
> Even UCL's early foundation in 1826 - as a secular institution as distinct from Oxford and Cambridge - set a precedent that was quickly followed...
> The internationalism, and the cosmopolitan character that it gives rise to, is a very special feature of UCL, both historically and even more so today... it is also an intellectual quality, of an institution that has a global vision, that sets out to address, through its teaching and research, problems which have universal significance.
> Malcolm Grant, President and Provost, 2004

Another excerpt from the beginning of the book states this:

> Unlike the modern civic universities which grew from some local patriotism, UCL grew from an idea. Originally that idea was based upon a belief in freedom of investigation without any distinction of creed or race or sex. That quest, which made University College the pioneer in modern university development in England, also led the founders to find a place in the College for studies which had previously been outside the university curriculum. The freedom to investigate was not an idle phrase: it meant breaking into new fields.
> Sir Ifor Evans[17] [43]

The current UCL web site sums it up this way: "UCL is one of the world's leading universities, founded in London to open up education to all on equal terms. Today our outstanding research and innovative teaching drive entrepreneurial solutions to the world's major problems... UCL is London's research powerhouse..." [44]. Much of this summary could apply equally to the University of California in the United States (absent the glaring continuing physical presence of Jeremy Bentham!), where David would spend the latter part of her career.

David told the story about how she got to UCL: "One day I was passing University College and I crashed my way in to see K Pearson. Somebody had told me about him, that he had done some actuarial work. I suppose it was just luck I happened to be there. Curious how fate takes one, you know. We hit it off rather well, and he was kind to me" [8]. This part of David's story reminds us of the ancient Greek expression that "character is your fate."

In fact, K Pearson had a reputation of being warm and friendly, a gentle person with a kind smile, and a person who was considerate of his students [9]. Nevertheless, many years later David described

---

[16] Note: In 2017 this number was 30.

[17] See the chapter UCL: A Survey, 1950–1955.

him as being "sometimes... really grumpy" and "rather pigheaded," and she said he was the only person who truly scared her [8]. But she also said she was "fond of him" and that the 2 years she spent with him were probably the hardest years she ever spent. She said that, in the projects that she worked on with K Pearson, she would do the mathematics because she was better at it than he was [10] (p. 133).

In David's time, the hub for statistics research at UCL was a building called the Bartlett Building, which later (in 1980) would become the Pearson Building [45].[18] It is located on Gower Street, the city street in front of the university's main gate, immediately to the left as you enter the gate. It is considered part of the Main Building, and it now houses the department of geography [46].

David was very young, around the age of 22, when she crashed her way in. Conversely, K Pearson was 74 and close to retirement, although he looked much younger than his age [9]. David recalled many years later that, in this first meeting with him, K Pearson asked whether she had a scholarship. When she replied "yes" because she had the Devon Major scholarship, he suggested he would get her scholarship renewed if she came to UCL to study [8]. David received permission to use the name of her principal at Bedford College as a reference for a postgraduate studentship at UCL to work on her PhD.[19]

David entered UCL, and her scholarship was renewed for another 2 years as a graduate student [47]. She moved into 1A Callow St. Chelsea, registered for the PhD in statistics to begin October 1931,[20,21] attended lectures toward a PhD in applied statistics (faculty of science), and did other things that graduate students do. She also calculated tables of correlation coefficients and worked on "Certain Theorems on Correlation," as you will see in the next chapter of this book.[22]

Many years later, David recalled that there was no plan discussed at the time of her meeting K Pearson that she would earn a PhD, but rather she should just "come and learn" [8]. However, the university records show that she entered UCL in 1931 as a PhD student, declaring her intention to complete her PhD in 1933.[23]

David entered a strong research environment at UCL. The department of applied statistics and eugenics, under K Pearson's direction, was much more of a research than teaching unit.[24] Programs were advertised as being adapted to the needs of postgraduates and "students training for special investigations."

Students could work toward a general or a special BSc degree. Fitting with the research interests in the department, honors students were encouraged to pair statistics with pure mathematics, anthropology, human anatomy, botany, psychology, or zoology. Courses were offered in the History of Statistics, Problems of Eugenics, Statistics for Medical Students, Vital Statistics, a Junior Course for those with little mathematical knowledge, and a Senior Course on the "whole modern theory of statistics." This Senior Course included association, contingency, correlation (basics, advanced theory, quantitative and qualitative categories, multiple correlation, applications), measurement of variation, theory of sampling and probable errors, interpolation, quadrature construction of tables, elements of finite differences, theory of probability, classical problems, applications to the modern theory of statistics, and the periodogram. Laboratory work was also offered. The Biometric Laboratory was

[18] Note: University College London has since denamed this building.

[19] Bedford College: handwritten note for the files, 1 page, November 1932.

[20] David, F.N.: University College London Student Registry File.

[21] David, F.N.: University College London First Entry Form.

[22] David, F.N.: University College London Re-Entry Form.

[23] David, F.N.: University College London First Entry Form.

[24] University College London: Report of the professorial board on the resignation of the Galton Chair by Professor K Pearson, APP. I. n.d.

reportedly well equipped (calculating machines, integrating mechanisms, moment-finders, statistical models, diagrams, measuring instruments, microscopes, cameras, memoirs, etc.).[25]

K Pearson was a progressive and broad thinker. As a student at the University of Cambridge in the United Kingdom, the University of Heidelberg, and the University of Berlin in Germany, he studied many diverse subjects: mathematics, physics, metaphysics, physiology, law, literature, and political science; he was even a published historian. Why did he study all of these subjects and more? He wrote this:

> I rush from science to philosophy, and from philosophy to our old friends the poets; and then, over-wearied by too much idealism, I fancy I become practical in returning to science. Have you ever attempted to conceive all there is in the world worth knowing - that not one subject in the universe is unworthy of study? The giants of literature, the mysteries of many-dimensional space, the attempts of Boltzmann and Crookes to penetrate Nature's very laboratory, the Kantian theory of the universe, and the latest discoveries in embryology, with their wonderful tales of the development of life - what an immensity beyond our grasp! [48]

K Pearson eventually settled into the mathematical sciences.

The story K Pearson told about his decision to work in the mathematical sciences is this: "Professor Beesly, just because I had lectured to revolutionary clubs, Professor Croom Robertson, just because I had written on Maimonides in his journal *Mind*, Professor Alexander Williamson, just because I had published a memoir on and criticized lectures of his on the Lake Poets, pressed me to be a candidate for the Chair of Mathematics!" K Pearson's interests were so broad that when he eventually retired in 1933, it took three people to replace him [43].

In 1884, K Pearson became the UCL professor of applied mathematics and mechanics. In 1903, he founded a Biometrics Laboratory, which existed through 1933.[26] He wanted "to make statistics a branch of applied mathematics with a technique and nomenclature of its own" and called it "biometry." He wanted it to be a serious branch of science and an exact science that required serious training, a "new calculus which was destined to become an element in the technique of many branches of science." K Pearson described this task to be a "very arduous one"—a "battle"—one that he fought for 20 years and eventually saw signs of winning [49].

As time went on, K Pearson became more and more interested in applied statistics—the relationship between sociological and biological data. He founded the department of applied statistics at UCL in 1911—the first statistics department in a university [50], "the first [department] of its kind."[27] UCB Statistician Elizabeth L. Scott called the department "...the birthplace of modern statistics. It was there that J. Neyman and E.S. Pearson collaborated on their famous series of joint papers... which are the very foundation of testing hypotheses."[28]

At the same time that he founded the department of applied statistics, K Pearson became the Galton professor of eugenics (1911–1933). Francis Galton had founded a Eugenics Laboratory and when he passed away, K Pearson was put in charge of it. This was in addition to being in charge of the Biometrics Laboratory ..."An enormous output of research continued on many aspects of intelligence, craniometry, heredity and 'questions of the day and of the fray'" [43, 51]. K Pearson specifically wanted to apply these ideas to help the human race. He became, in fact, obsessed with the ideas of Francis Galton and Galton's cousin Charles Darwin about natural inheritance and eugenics. This was an obsession that would eventually put a tarnish on K Pearson's legacy.

An interesting connection to this story is Florence Nightingale. According to K Pearson:

---

[25] University College London Calendar 1931–1932.

[26] Pearson, E.: letter to Haldane, J.B.S. 12 March 1948.

[27] Typed manuscript about Pearson E. labeled "N.L. Johnson in Frequency Curves," 9 pages, n.d.

[28] Scott, E.L.: letter to Knight, W.D. 11 March 1969. Note: Walter D. Knight was dean of the UCB College of Letters and Science.

> The idea of a Professorship of Statistics to be founded in an English University appears to have first occurred to Florence Nightingale. She desired to found a professorship of applied statistics and she consulted [Benjamin] Jowett and Francis Galton on the subject. The matter did not come to fruition because the funds she could afford to appropriate to the foundation were inadequate for the purpose in the opinion of her advisors. She asserted that most social problems failed of their fitting solutions because there was no adequate training in statistical science, and went so far as to assert that man could only ascertain the mind and purpose of the Deity by studying his laws as exhibited statistically. One side and one side only of Florence Nightingale's work has been commemorated in the funds raised for the training of nurses. The future will probably show that the other dominant idea of her mind, the training of statisticians, was from the national standpoint of equal, if not greater, importance. No finer memorial could be raised to her memory than the endowment of a laboratory to fulfil this function, and the association of this endowment with her name. The suggestion that the combined department of the Biometric and Galton Laboratories should be called the Department of Applied Statistics was an echo of Florence Nightingale's plea for a professorship of Applied Statistics.

Galton extended Nightingale's vision by widening her concept of applied statistics.[29]

David would go on to work in the department of applied statistics at UCL. She would take after her distinguished mentor K Pearson and become one of the broadest statisticians of her time. Twenty years after her first meeting with K Pearson, J.B.S. Haldane would commend David for her breadth: "... a number of people with Fellowships of the Royal Society and the like are narrow specialists compared with Dr. David..."[30]

## References

1. Wild, W.: c. 1900s–1950s: Bad girls on bikes: In the early 20th century, these women enjoyed a good ride (2021) https://www.considerable.com/entertainment/retronaut/female-motorcycle-riders/ Accessed 2 September 2021
2. Vintage Everyday: 22 amazing vintage photographs of women riding motorcycles from the 1920s and 1930s 11 March (2014) https://www.vintag.es/2014/03/vintage-photographs-of-women-and.html Accessed 2 September 2021
3. Champneys, B.: The Victorian Web: Bedford College for Women, London (1942-1935) (2017) http://www.victorianweb.org/art/architecture/champneys/7.html Accessed 12 May 2017
4. Bentley, L.: Educating Women: A Pictorial History of Bedford College University of London 1849–1985. Alma Publishers, Surrey (1991)
5. UCL Bloomsbury Project: Ladies' College (2017) http://www.ucl.ac.uk/bloomsbury-project/institutions/ladies_college.htm Accessed 12 May 2017
6. Wikipedia: Royal Holloway College, University of London (2017) https://en.wikipedia.org/wiki/Royal_Holloway,_University_of_London#Location] Accessed 12 May 2017
7. Wikipedia: Bedford College, London (2017) https://en.wikipedia.org/wiki/Bedford_College,_London Accessed 12 May 2017
8. Laird, N.M.: A conversation with F.N. David. Statistical Science 4(3), 235–246 (1989)
9. Stouffer, S.A.: K Pearson—An appreciation on the 100th anniversary of his birth. Journal of the American Statistical Association 53(281), 23–27 March (1958)
10. Reid, C.: Neyman - From Life. Springer-Verlag, New York (1982)
11. Lehmann, E.L.: Reminiscenses of a Statistician: The Company I Kept. Springer (2008)
12. Hafeman, M.: The role of the actuary in insurance, Primer Series on Insurance, Issue 4, Washington DC, The World Bank, May (2009) http://siteresources.worldbank.org/EXTFINANCIALSECTOR/Resources/282884-1242281415644/Role_of_Actuary_in_Insurance.pdf Accessed 21 August 2017
13. International Actuarial Association: The role of the actuary. June (2013) http://www.actuaries.org/LIBRARY/Papers/Role_Actuary_EN.pdf Accessed 21 August 2017
14. Hellwig, T.T.: The origins of unemployment insurance in Britain. Social Science History 29(1), 107–136 (2005)
15. Garside, W.R.: British Unemployment 1919–1939: A Study in Public Policy. Cambridge U.P. p. 37 (2002)
16. Mowat, C.L.: Britain Between the Wars: 1918-1940, pp. 43–46 (1955)
17. Laybourn, K.: Modern Britain Since 1906: A Reader. Tauris History Readers. I.B. Tauris, p. 105 (1999)

[29] Pearson K.: History of the Biometric and Galton Laboratories, University of London, unpublished manuscript (1920).
[30] Haldane, J.B.S.: letter to Pearson, E. 18 July 1957.

18. Deeply Trivial: Great minds in statistics: F.N. David versus the patriarchy 8/32 (2017) http://www.deeplytrivial.com/2017/08/great-minds-in-statistics-fn-david.html Accessed 20 December 2022
19. Hickman, J.: History of the actuarial profession. Encyclopedia of Actuarial Science (2017) John Wiley and Sons (2004) https://www.wiley.com/legacy/wileychi/eoas/pdfs/TAH012-.pdf Accessed 23 August 2017
20. Morgan, K.: Women in the profession. The Actuary: The Magazine of the Institute and Faculty of Actuaries 2/20 (2013) http://www.theactuary.com/features/2013/02/women-in-the-profession/ Accessed 29 August 2017
21. University College London: K Pearson (2017) https://www.ucl.ac.uk/statistics/department/pearson Accessed 13 September 2017
22. Pearson, K.: On the criterion that a given system of deviations from the probable in the case of a correlated system of variables is such that it can be reasonably supposed to have arisen from random sampling. Philosophical Magazine Series 5, 50(302), 157–175 (1900)
23. Pearson, K.: On lines and planes of closest fit to systems of points in space. Philosophical Magazine Series 6, 2(11), 559–572 (1901)
24. Jolliffe, I.T.: Principal Component Analysis 2nd ed. Springer-Verlag, New York (2002)
25. Pearson, K.: Contributions to the mathematical theory of evolution. II. Skew variation in homogeneous material. Philosophical Transactions of the Royal Society A (Mathematical, Physical and Engineering Sciences) 186, 343–414 (1895)
26. Holland, E.: The Men and Women's Club. Edwardian Promenade 13 December (2009) http://www.edwardianpromenade.com/love/the-men-and-womens-club/ Accessed 23 August 2017
27. Love, R.: Alice in eugenics-land: Feminism and eugenics in the scientific careers of Alice Lee and Ethel Elderton. Annals of Science 36(2), 145–158 (1979)
28. Porter, T.M.: K Pearson: British mathematician. Encyclopedia Britannica (2017) https://www.britannica.com/biography/Karl-Pearson Accessed 7 September 2017
29. Schreiner, O.: Women and Labour. London (1911)
30. Psychology's feminist voices: Profile: Alice Lee (2017) http://www.feministvoices.com/alice-lee/ Accessed 7 September 2017
31. Elderton, E.M.: A first study of the influence of parental alcoholism on the physique and ability of the offspring, Francis Galton laboratory for national eugenics, University of London. Eugenics Laboratory Memoirs, v.X (1910)
32. The European Mathematical Society: Pearson, Karl. Encyclopedia of Mathematics. Springer (2017) https://www.encyclopediaofmath.org/index.php/Pearson,_Karl Accessed 7 September 2017
33. University College London Department of Statistical Science: K Pearson (2017) https://www.ucl.ac.uk/statistics/department/pearson Accessed 7 September /2017
34. Institute and Faculty of Actuaries: The Actuaries' Club (2018) https://www.actuaries.org.uk/get-involved/dining-clubs/actuaries-club Accessed 12 June 2018
35. Wikipedia: William Palin Elderton (2017) https://en.wikipedia.org/wiki/William_Palin_Elderton Accessed 13 September 2017
36. Pearson, K.: History of the Biometric and Galton Laboratories. University of London (1920)
37. Denby, L., and Landwehr, J.: A conversation with Colin L. Mallows. International Statistical Review 81(3), 338–360 (2013)
38. Barton, D.: Obituary: Florence Nightingale David. Journal of the Royal Statistical Society Series A (Statistics in Society) 157(2), 299–301 (1994)
39. Internet Encyclopedia of Philosophy: Jeremy Bentham (1748-1832) (2017) http://www.iep.utm.edu/bentham/ Accessed 5 September 2017
40. Oxford Living Dictionaries: Auto-icon (2017) https://en.oxforddictionaries.com/definition/us/auto-icon Accessed 6 September 2017
41. Meier, A.: Skeleton of 18th-Century philosopher Jeremy Bentham will travel to the Metropolitan Museum. Hyperallergic (2018) https://hyperallergic.com/415209/jeremy-bentham-met-breuer/ Accessed 5 September 2018
42. University College London: Nobel prize winners (2017) http://www.ucl.ac.uk/about/who/history/nobel-prize-winners Accessed 5 September 2017
43. Harte, N. and North, J.: The World of UCL 1828-2004, Third Edition. University College London Press, London (2004)
44. University College London: University College London (2017) http://www.ucl.ac.uk Accessed 11 September 2017
45. University College London: UCL denames buildings named after eugenicists 19 June (2020) https://www.ucl.ac.uk/news/2020/jun/ucl-denames-buildings-named-after-eugenicists Accessed 27 March 2023
46. University College London: Map highlighting Pearson Building and 26 Bedford Way (2017) http://www.geog.ucl.ac.uk/about-us/contact-us/downloadable-maps/ucl-map-highlighting-pearson-building-and-26-bedford-way/image_view_fullscreen Accessed 20 September 2017
47. Pulman's Weekly News: Miss F.N. David's success. 9 October (1935)
48. Pearson, K.: The new werther. C, Kegan Paul and Co., London, p. 6 (1880)

49. Pearson, E. (presumed): unpublished history manuscript titled University of London, History of the Biometric and Galton Laboratories, 1920.
50. University College London: Department of Statistical Science, K Pearson (2018) https://www.ucl.ac.uk/statistics/department/pearson Accessed 1 June 2018
51. Porter T.M.: K Pearson: The Scientific Life in a Statistical Age. Princeton, New Jersey: Princeton University Press (2004)

# Part III

# In the Trenches

# Rising (1931–1938) 4

*Here will I set up my everlasting rest, and shake the yoke of inauspicious stars from this world-wearied flesh.*

Shakespeare, W.: Romeo and Juliet (1597), quoted by David, F.N. on a draft manuscript (around 1932)

## 4.1 Putting Her Name Forward

David dropped into an environment that had a rich legacy. Statistics at UCL began as the Biometric Laboratory in 1895. It was at the time K Pearson (then professor of applied mathematics and mechanics) gave the first course on mathematical statistics to two or three postgraduate students, "probably the first course given on the modern mathematical theory [of statistics] at all." The course was seminal: It was given every year thereafter, and one of the first students became reader of statistics at Cambridge University.

E Pearson (Egon) recalled the aims of his father and the struggles of the time:

> ...as the field of this form of investigation had been very little worked, a school sprung up which has since been recognized as the "Biometric School," and the group of workers, occupying a small room at University College, later termed the Biometric Laboratory, issued a long series of memoirs, which formed the basis of the English school of mathematical statistics. The object of this school was to make statistics a branch of applied mathematics with a technique and nomenclature of its own, to train statisticians as men of science, to extend, discard or justify the meagre processes of the older school of political and social statisticians, and in general to convert statistics in this country from being the playing field of *dilettanti* and controversialists into a serious branch of science, which no man could use effectively without adequate training, any more than he could attempt to use the differential calculus, being ignorant of mathematics. This task was a very arduous one, for statistics in one form or another are fundamental in nearly every branch of science in precisely the same manner as mathematics are fundamental in astronomy and physics. Inadequate and even erroneous processes in medicine, in anthropology, in craniometry, in psychology, in criminology, in biology, in sociology, had to be criticised, not for the pleasure of controversy, but with the aim of providing those sciences with a new and stronger technique. The battle has lasted for nearly twenty years, but there are many signs now that the old hostility is over and the new methods are being everywhere accepted.[1]

This was 10 years before David arrived on the scene and collaborated to convert statistics to an essential branch of science which men *or women* could use effectively only with adequate training.

[1] Pearson, E. (presumed): unpublished history manuscript titled University of London, History of the Biometric and Galton Laboratories, 1920.

 
A. L. Golbeck, *Florence Nightingale David*,
https://doi.org/10.1007/978-3-031-77768-4_4

David's role over the course of her career in combatting the kind of sexism that is inherent in E Pearson's words cannot be dismissed.

K Pearson named the new science *Biometry*. Then he received a grant through the university, from the Worshipful Company of Drapers (a company founded in the Middle Ages to make uniforms), to support the work of the department through such mechanisms as research assistants and the publication of the *Drapers' Company Research Memoirs*. The Drapers continued to fund Biometry at UCL for many years.

K Pearson was a massively important player at UCL. The history book for UCL [1] says that when he retired it took three people to fill his shoes.[2] By 1931, K Pearson had built an academic department that was centered on research. The University Calendar[3] said this about the department of applied statistics and eugenics: "This Department is largely a research department, adapted to the needs of post-graduate workers and of students training for special investigations. Results reached in the Laboratories are occasionally dealt with in more popular courses of lectures."

In addition to all of his other interests, K Pearson was very interested in history and offered lectures on, among other things, the History of Statistics and The French Statisticians and writers on Probability: Duvillard, Poisson, and recent writers.[4] Young David shared K Pearson's interest in history.

David affiliated with a statistics professional association early on in her career. This was with the Royal Statistical Society, founded in 1834. In David's time, this was largely a man's society. In 1858, Florence Nightingale became its first woman member. It is ironic that it was David's namesake who achieved this. Furthermore, the Society did not have a woman president until 1975.

Also in David's time, the Royal Statistical Society had discussion meetings which included, among other things, deliberations of current work. There were two kinds of persons who were allowed to attend such meetings: *Fellows* and other persons called *associates*. David became an associate in February 1936, but a November 1937 ledger indicated she never paid her dues (about 10 shillings).[5,6]

David became a Fellow of the Society in May 1948,[7] She was a member of the research section in 1951–1952 and the council in 1959–1961.[8] She kept her membership after moving to the United States in the late 1960s.

David received a number of mentions in the statistics literature for her work during her years as a young scholar. One was a paper by Frank Garwood on correlation coefficients. Garwood studied statistics at UCL and later worked with David on her war work. Garwood knew that David was computing probability integral tables for the distribution of the correlation coefficient. In his article, he notified readers of David's work on correlation coefficients and identified which formulas she was using for small and large samples [2].

Another early mention of David was by K Pearson himself. This was in his article, "On a method of determining whether a sample of size n supposed to have been drawn from a parent population having a known probability integral has probably been drawn at random." First he referred to David's

[2] Note: One of the three people who replaced K Pearson was his son Egon, who worked closely with Jerik (aka Jerzy) Neyman. Another was J.B.S. Haldane, who became a friend to David, F.N. The third was R.A. Fisher.

[3] Note: In the United States, University Calendars are known as University Academic Catalogs.

[4] University College London Calendar 1931–1932.

[5] Royal Statistical Society: proposed discussion meetings, 4 June 1928. Note: draft of proposal to be laid before the meeting on Monday, 4 June 1928.

[6] Royal Statistical Society: notices of study group meetings, November 1937.

[7] Royal Statistical Society: nomination paper of David, F.N. May 1948. Note: Peter Armitage (b.1924–d.2024) became a Fellow of the Royal Statistical Society the same year, in February 1948.

[8] Royal Statistical Society: council meeting minutes, 12 March 1959; 15 October 1959; 12 November 1959; 10 December 1959.

work that had already been published with himself and Stauffer. It was on the Bessel function. Then, like Garwood, K Pearson announced David's work on the tables of the correlation coefficient which would be published soon. Finally, he acknowledged David's computing work on the Incomplete $\beta$- and $\Gamma$-Function Tables [3].

David had become valuable to K Pearson. But K Pearson was preparing for retirement. He let it be known to his advisory board that he intended to retire from the position as Galton Chair at the close of the 1932–1933 academic year.[9] He had served the university as a professor for 49 years, including as the Galton professor for 21 years.[10] Facing a significant organizational transition, David's uncertain future became more uncertain, just as the international political situation in Europe and Asia was becoming more uncertain.

In September 1933, after the completion of her scholarship and following his retirement, K Pearson hired David as his research assistant. This was in the government department of industrial and scientific research. She started in October [4].[11] The department was set up in response to conditions of World War I where there was considered to be too much reliance on foreign industries. The department cooperated with universities and was responsible for encouraging, organizing, and developing research.[12] David had become the research assistant to "the world's greatest statistician" [4].

David was to help K Pearson with his research and with his editing of the journal *Biometrika*. In the beginning, it wasn't easy to publish mathematical statistics research, it being a new branch of science. Accordingly, W.F.R. Weldon (b.1860–d.1906) suggested that a special journal be established. K Pearson named it *Biometrika*,[13] and he edited it with Weldon and Francis Galton (Fig. 4.1).

David described her role with the journal: "*Biometrika* papers would come in, and [K Pearson] would not quite see the point so he would rewrite them. I would do the mathematics. I was a little better at that than he was" [5] (p. 133). At the time this book was written, *Biometrika* continued to do well, with an impact factor of 2.445, putting it in the top 40 percent of journals. Produced by Oxford University Press, it includes "papers containing original theoretical contributions of direct or potential value in applications" [6].

David also reportedly did a little statistics work for Lord Bertrand Edward Dawson (b.1864–d.1945), who at the time was King George V's doctor. She was still living in London near Jubilee Place in Chelsea [7].[14]

When K Pearson retired in 1933, UCL split the department into two, and two heads were hired, one for statistics and one for eugenics. In 1937, a department of biometry was also formed. Ten years later, K Pearson's material holdings (i.e., archives) were still being sorted among the three departments (statistics, eugenics, and biometry).[15]

One of the new heads was K Pearson's son Egon (E Pearson), who took over statistics and for a few years thereafter referred to his department as the "New Department of Statistics." There were a lot of things to manage in the split, including the relocation of miscellaneous properties, large and

---

[9] University College London: Galton subcommittee agenda, 19 June 1933.

[10] University College London: report of the professorial board on the resignation of the Galton Chair by Professor K Pearson, 19 June 1933.

[11] Bedford College: student record of David, F.N. n.d.

[12] Note: The department of industrial and scientific research was dissolved in 1965.

[13] Pearson, E.:. History of the Biometric and Galton Laboratories. University of London, five pages unpublished 1920.

[14] Bedford College: handwritten notes "Old Students," n.d. Note: Dawson became president of the Royal College of Physicians (1931–1937). In 1936, in a controversial action that only became known many years after Dawson's death, Dawson euthanized King George V. Dawson maintained this was to allow the king to die in dignity, whereas some saw it as a "convenience killing."

[15] Pearson, E.: letter to Haldane, J.B.S. 12 March 1948.

**Fig. 4.1** Golbeck, A. at the UCL statistics department library, with the first volume of *Biometrika* (2017) [Image credit: Craig A. Molgaard]

small. For example, David had in her room six pictures of members of the Galton family.[16] David lost those pictures, as R.A. Fisher wanted them in his eugenics department.[17] There was a lot of haggling between Fisher and E Pearson prior to the finalization of their appointments.

E Pearson earned his bachelor's degree in 1920. The next year he was hired by his father into an assistant position in the department of applied statistics. By 1933 he had earned the rank of senior lecturer. He served as assistant editor of *Biometrika* starting in 1925. He earned his DSc degree the next year for work that he had published. He had about 25 publications, a number of which were with Neyman. When E Pearson took over as head of the new department of statistics, he was promoted from the rank of senior lecturer to the rank of reader.

At the same time, Edith Elderton's title was changed from reader in eugenics to reader in social statistics.[18] Even though Elderton had a higher academic rank than E Pearson, and they both had DSc degrees, he was given the head of the new department. Nevertheless, the provost reported to E Pearson that she was "entirely agreeable" to work under him, at least when it had been discussed "at a much earlier stage," and that "she herself volunteered the proposal—there was no hinting at the suggestion on my part."[19]

E Pearson felt compelled to justify the fact that Elderton was passed over. He wrote:

> As long as she remains at University College, Miss Elderton will naturally be second in command of the Department, but her position as a University Reader is in any case on a special basis. She is not concerned with the more mathematical side of the teaching, and I should like to be able straight away to secure and put into training the suitable right-hand man for the mathematical position. He would be needed, for example, to take his place as soon as possible as a recognized teacher on the Board of Studies in Statistics, which with the loss

[16] Pearson, E.: letter to Fisher, R.A. 29 May 1936.

[17] Kern, M.M.: letter to Pearson, E. 10 July 1936.

[18] University College London: Galton subcommittee meeting agenda, 19 June 1933.

[19] Mawer, A.: letter to Pearson, E. 24 April 1933.

of my father and Dr. Stocks will be badly under-represented by University College; further he will be needed as University Examiner.[20]

A right-hand man. He, his, he. Elderton was passed over, and then she was minimized.

Also at the same time that E Pearson was made head of the statistics department, E.C. Fieller (b.1907–d.1960) was promoted from mathematical assistant (his title under K Pearson) to lecturer. Fieller later became known for Fieller's theorem, which allows a confidence interval to be calculated for the ratio of two means.

At his promotion to department head, E Pearson was cited for his 5-month sabbatical to the United States in 1931, where he formed a relationship with W.A. Shewhart at Bell Labs in New York, "the man who has been largely responsible for raising in the United States the question of the introduction of statistical methods in the control of quality in industry." E Pearson brought Shewhart to London to lecture, which resulted in a number of developments, including a new section of the Royal Statistical Society on industrial and agricultural research, and including advising the government on problems related to industrial standardization and specification, including sampling methods and testing.[21] When he assumed the department head of statistics, E Pearson had to look in new directions for "raw material," and he saw industrial applications of statistics as a promising direction. He fruitfully pursued this direction for many years.

E Pearson understood that the value of a statistics department is more than the number of undergraduates taught and that its real value is in the research one produces. This followed the preferences of Sir Francis Galton and K Pearson who preceded him.[22] Nevertheless, he noted that the statistics department enrollments increased rapidly after 1931. Many students were coming from abroad. By 1933 there were 12 postgraduate students, and by 1945 the department was looking for more undergraduate students. The UCL Calendar stated: "[The general course in statistics] aims at providing students with the statistical tools required in a wide variety of practical problems, ranging from the biometric and sociological to the engineering and industrial. While the approach is a mathematical one, special emphasis is laid on the statistical craftsmanship needed to put theory into application."[23] The students would start out by working on problems given in classes, and then later they would be given individual problems to solve based on their aptitudes and interests.[24] In his 1934–1935 report about the statistics department, E Pearson wrote:

There is evidence of an increasing demand for the competent trained statistician, and this year statistical posts have come to my notice which could not be filled from university College because the right kind of man was not available. . . The acceptance of a new technique with unfamiliar conceptions and terminology is naturally a slow process, and the soundest course, in the long run, is to aim at interesting the younger men.[25]

Today, we find E Pearson's gendered language to be jarring.

By 1935, E Pearson thought his department had become well balanced, the research culture for the students was good, and the number of students had increased slightly. The department was now offering a 2-year MSc program in statistics, which was of interest to mathematics students.[26]

David was a part-time assistant to K Pearson, and in 1935 she was about to make her first move up the ladder, thanks to the efforts of E Pearson in the early part of his tenure as head of the department of

[20] Pearson, E.: letter to provost, 17 May 1933.

[21] University College London Professorial Board: document detailing E. Pearson's appointment, 6 March 1935.

[22] University College London Professorial Board: report on the resignation of the Galton Chair by Professor K Pearson, Appendix 1, December 1932.

[23] University College London Calendar 1933–1934.

[24] Pearson, E.: report of the statistics department session 1933–1934.

[25] Pearson, E.: report of the statistics department session 1934–1935.

[26] Pearson, E.: report of the statistics department session 1935–1936.

statistics. E Pearson had several key priorities in March 1935. One was to get himself promoted from a reader rank to a professorship, which happened (effective 1 October 1935). Another was to appoint Neyman for the next academic year and to appoint David as a full-time assistant in the department starting in the next academic year. These appointments would maintain the number of staff members in the department at four. David "has been working with my father [K Pearson], but wishes to gain rather wider experience both in teaching and research," E Pearson reported.[27,28] E Pearson arranged for David's salary to double.[29]

E Pearson put a stack of correspondence in his file that he stapled together with a note on top: "Oct 1935–Feb 1938, when I was trying to get a better salary for F.N." The department was short on funds because the funds had to be used to pay an increase in his salary when he was appointed professor and head of the department. He tried creative financing: Would the provost allow him to put money that he received from his evening lectures into the department grant so that he could increase David's salary?

The department was already employing K. Pearson, E. Elderton, E. Pearson, E.C. Fieller, M.S. Bartlett, J. Neyman, and B.L. Welch. But David was also needed:

> It would be impossible to carry out the necessary work this year without using Miss David as a full time Assistant, unless opportunity for research by myself, Dr. Neyman and Mr. Welch were to be very seriously cut down. In a Department containing such a large proportion of postgraduate workers, this would be most undesirable, as problems for research for these students can only be found if members of the staff are themselves able to push forward with their own work and ideas. Consequently I have seen no alternative but to offer Miss David 200 GBP a year, realizing that the extra 100 GBP could be obtained from the departmental grant.[30]

The university finance secretary and accountant approved the amount of the increase, as did the college committee, but the accountant made different provisions about how it would be paid.[31,32]

K Pearson passed away on 27 April 1936. E Pearson reported that his father was:

> ... the founder of modern statistics and the great majority of mathematical statisticians alive in England today and many of those scattered over the world, received their trainings from him at University College. Material traces of this history of statistical teaching which has been so closely associated with the history of the development over forty five years of a great [area] are to be found in numberless places in the building containing the present Department of Statistics: books, models, pictures, instruments, statistical records and material. We shall now have back with us also in the shape of a special 'K.P.' Library, the whole of his collection of statistical and mathematical books, journals, pamphlets, offprints and unpublished lecture notes which with his collection of pictures and busts, he wished should be used "to the advantage and pleasure of the students and staff" of this Department of University College.

E Pearson also noted that the journal *Biometrika* would once again be managed by the statistics department.[33]

David's first statistics mentor was gone.

---

[27] Pearson, E.: Department of applied statistics, 28 March 1935.

[28] University College London Academic Registrar: letter to Pearson, E. 18 July 1935.

[29] University College London Secretary: letter to Pearson, E. 4 December 1935.

[30] Pearson, E.: letter to UCL provost, 15 October 1935.

[31] University College London Finance Secretary and Accountant: letter to Pearson, E. 6 November 1935. Note: It is unclear what these different provisions were.

[32] University College London College Committee Secretary: letter to Pearson, E. 4 December 1935.

[33] Pearson, E.: report of the statistics department session 1936–1937.

## 4.2 The Neyman Years

David's next statistics mentor was Jerzy Neyman. The Neyman years were 1934–1938. David described these years in this way: "Most of the time I was babysitting for Neyman, explaining to the students what the hell he was up to" [5] (p. 133).

E Pearson hired Neyman into his department of statistics, as an "assistant" to teach postgraduates. Neyman was hired with the title of Special Lecturer, starting at the third term (summer) of the 1933–1934 session, and with the same salary as Ethel Elderton.[34] This was the year after E Pearson assumed the position of chair of the statistics department.[35]

Neyman had already spent some time in the department, having been there at the invitation of K Pearson on a Rockefeller Fellowship in 1925–1926, during which time he was stimulated by E Pearson's interest in "the three cornered dispute on foundations of statistics" mainly involving K Pearson, E. Edgeworth, and R.A. Fisher. So Neyman was familiar with the operations of the department and university, and he self-described his own "work on serious problems" as having begun in 1926 after this year of fellowship at UCL.[36,37]

Neyman was thrilled to be going back to London. He sent a telegraph to E Pearson from Warsaw: "OK HIP HIP HURRA THANKS = JUREK."[38] Neyman wrote an accompanying letter to E Pearson: "I have just been in a wave of pessimism and depression when your letter arrived. What I am thinking at present will tell you my wire."[39] It was a letter of negotiation. Neyman explained how hard it was for his current employer, University of Warsaw, to release him in the middle of a session and how it would be easier to get permission if his title didn't include the word "assistant."

To say that Neyman was interested in going to London would be an understatement. He told E Pearson that he would "jump at once for as long as I could have money to live on... Oh, Egon you do not imagine what your proposal means to me. I am nearly crazy."[40] E Pearson followed up by preparing two letters, one without the word assistant which would be presented to the Polish Ministry of Education and the other with the word assistant so that Neyman would be clear that he would be acting as an assistant in the department.

E. Pearson assured his provost that Neyman would be willing to do anything he was asked, even to stay on after his term ended. Then E Pearson asked the provost for a larger salary than the provost had indicated: "On the one hand I know that owing to the desperately hard problem of making ends meet which confronts many University lecturers in Poland, he [Neyman] would be ready to come over here into easier conditions for teaching and research, for any sum which is enough for him and his wife to live on. On the other hand, he is a personal friend of mine, and I do not want to be urging the College to pay more than is reasonable."[41]

The letter which invited Neyman gives interesting information and perspectives about the state of the field of statistics at the time:

> The lectures would be given to postgraduate students, and I would like you to deal specially with that important field of statistics, still so little studied by the writers of textbooks to which you have made many contributions. [para] A clear statement of the underlying principles on which the methods of statistical inference should be

---

[34] Pearson, E.: report of the statistics department session 1933–1934.

[35] University College London Calendar 1934–1935.

[36] Neyman, J.: handwritten autobiographical data, eight pages, 8 July 1963.

[37] Pearson, E.: letter to Neyman, J. 30 January 1934.

[38] Neyman, J.: post office telegraph to Pearson, E. 27 January 1934.

[39] Neyman, J.: letter to Pearson, E. 27 January 1934.

[40] Neyman, J.: letter to Pearson, E. 27 January 1934.

[41] Pearson, E.: letter to the University College London provost, 30 January 1934.

> based, is not only of importance for the better development of theory but tends, as you know, to train the mind in clear thinking when faced with practical problems such as those that are arising in the application of statistical methods in industry. I feel sure that you will be able to present the subject in a way which will combine the mathematical theory with illustrations of its practical importance... I hope very much that it will be possible for the University authorities in Warsaw to grant you leave of absence, and so to give us the opportunity of introducing into our teaching in London a new outlook and personality which is specially valuable to a Department which collects students from many quarters of the world.[42]

As it turned out, it wasn't Neyman who would write those textbooks. It would be David.

Neyman came, and then he continued for the 1934–1935 session. During that session, Neyman gave a "very complete" set of advanced lectures on his contributions to the theory of statistical inference, and he directed the research of several students. The UCL Calendar listed him as teaching a first year course in elementary probability, two second year courses (one on advanced probability and the other on elementary theories of estimation), and two postgraduate courses (one on theories of estimation and testing statistical hypotheses and the other a follow-up course on the theory of characteristic functions). He also taught with E Pearson a seminar on "problems of statistical interest under investigation by workers in the department."[43] David was not yet teaching. 1935–1936 was the first year that she was listed in the UCL Calendar. She was listed as an assistant. E Pearson had just been promoted to professor and Neyman to reader in statistics.[44]

Ethel Elderton retired around this time, but she would remain in the department in an honorary position. She had worked for the college and department for 30 years and was a very generous and encouraging person. E Pearson noted that it was her who first taught others how to use a calculating machine.[45]

In 1935, Neyman was given a permanent position as reader in statistics. Neyman and E Pearson were continuing their work on the mathematical theory of statistics, only at a more advanced level than previously. E Pearson reported to his university administration that the work was progressing well, "...largely owing to the stimulus and energy of Dr. Neyman."[46] K Pearson had started the *Drapers' Company Research Memoirs* in 1904. E Pearson decided to follow in his father's footsteps and also start a journal with his department, which he named *Statistical Research Memoirs*. The plan was to publish much of the Neyman-Pearson research in this journal. E Pearson described Neyman's contributions to the department as many and varied, and that these contributions were recognized in the form of two invitations. One was to France to give a six-lecture course at the Institut Henri Poincaré in Paris. The other was to the United Kingdom to give a paper at the Economic Society Congress that would be held in Oxford.

David had an appendectomy in March 1936 at University College Hospital in London. K Pearson warned her that doctors could make mistakes, expressed the hope that the operation would remove "the course of much trouble in the past," counseled her to take 3 or 4 weeks to rest after the operation, and wished her a speedy recovery.[47] David, of course, did not heed K Pearson's advice about the 3 weeks of rest and returned to work earlier, only to be scolded by him: "I do not think you are right in going back to work, before..." and telling her that they already had "plenty to go on with." "Please get away as soon as you can and do no work for three weeks," he lectured her.[48]

---

[42] Pearson, E.: letter to Neyman, J. 31 January 1934. Note: This letter was presumably a draft written by E Pearson to send to the provost.

[43] University College London Calendar 1934–1935.

[44] University College London Calendar 1935–1936.

[45] Pearson, E.: report of the statistics department session 1934–1935.

[46] Pearson, E.: report of the statistics department session 1935–1936.

[47] Pearson, K.: handwritten letter to David, F.N. 19 March 1936.

[48] Pearson, K.: handwritten letter to David, F.N. 3 April 1936.

By 1936–1937, there were 11 undergraduates and 23 postgraduates (18 from overseas) studying in the UCL statistics department. David contributed to this growth. The field was so new that E Pearson expressed concerns about being able to place students into good jobs. He added, with vision: "While we have little or no responsibility for the placing of the Dominion and foreign research students, it is satisfactory to feel that we are helping in the training of young people, many of whom in turn will start teaching and developing a new subject in Universities widely scattered throughout the world." This was also the year that E Pearson started to give an introductory course at UCL on statistical methods and probability. To his surprise, 140 people attended the first lecture. This suggested to him "that regular lectures of this type should become a permanent feature of the Statistics Department."

David in 1937 was living at 407 Addison House, a stately five-story brick apartment complex.[49] E Pearson reported that David "has proved herself an excellent teacher and thrown herself unstintingly into the junior work. She has been completing a lengthy piece of computation commenced under K Pearson's direction for a Table to be published shortly."[50]

This work on correlation would become David's first book. Her confidence interval charts were especially noted.[51]

In 1936–1937, Neyman gave six lectures on "Statistical Methods in Social and Economic Investigations." He was assisted by David and another person in the department. The lectures were attended by a mixture of people in business/industry and government. At this time, Neyman also gave "a very strenuous" series of lectures on mathematical statistics at a number of universities in the Eastern and Midwestern United States over a 6-week period. He was also invited to Switzerland to attend a very small (20 persons) International Congress on the Theory of Probability.[52]

It didn't take long for Neyman's salary to become an issue. In 1938, E Pearson made an attempt to raise it, but told his provost that it should "... not be in any way prejudicial to the prospects at future dates of other members of the Staff of the Department, i.e., Dr. Welch and Miss David" and "... the raising of Dr. Neyman's salary now must not be regarded as preventing him receiving further increments from time to time in the ordinary way."[53] Thus, E Pearson was looking out for Neyman, but not at the expense of David or another.

E Pearson had a very close collaboration with Neyman. They were academic partners. E Pearson told his provost: "In the last ten years Dr. Neyman and I have developed what may be described as a new method of introducing the theory of probability into statistical problems. In this our contributions have been complementary." E Pearson then talked about their "considerable" joint papers. Neyman filled a niche in the department that E Pearson didn't think he could fill by anyone else. In his view, Neyman was needed to teach advanced topics in the mathematical theory of probability. He wrote: "This subject has received relatively little attention... in England. It may be noted that last year Cambridge University, having no specialized teacher of its own in the subject, turned to Professor H. Cramer of Stockholm to write the pamphlet on Probability in its mathematical series. Neyman's approach has much in common with Cramer's."

He further wrote: "With the solution of the theory underlying the more essential practical methods, advanced work in the theory of statistics is becoming inevitably more and more mathematics. Of the postgraduate students there are always a number with a first rate training in modern mathematics; such workers are competent and anxious to deal with research problems needing in their solution mathematical tools with which I myself am not fully familiar." Neyman, as he saw it, could keep up

---

[49] David, F.N.: letter to Miss Haydon, 17 June 1937.

[50] Pearson, E.: report of the statistics department session 1936–1937.

[51] Pearson, E.: handwritten letter to Neyman, J. 28 August 1968.

[52] Pearson, E.: report of the statistics department session 1936–1937.

[53] Pearson, E.: letter to University College London provost, 21 February 1938.

with both the statistics literature and the mathematics literature. He "... has sufficient experience and originality of mind to supervise the work of a number of postgraduate students (among whom are sometimes included foreign professors of mathematics)."

Also, Neyman's international reputation was growing. While away lecturing in the United States, he received two professorship offers at about 4,500 USD, which is about 87,309 in 2021 USD. At London, he was receiving 550 GBP, which is about 37,858 in 2020 GBP. E Pearson described the differences as "very large" and told his provost that he couldn't advise "... a man of [Neyman's] standing to so continue." Neyman didn't want to leave the United Kingdom, and E Pearson hoped, and deemed it desirable, for Neyman to remain.[54] But Neyman did not.

E Pearson expressed his sadness at losing Neyman. He wrote to his colleague Walter Shewhart, expert on industrial quality control:

> ... in spite of many and long-drawn out efforts, it has been impossible to get University College to advance or give any hopes of advancing [Neyman's] salary... In London there is no elasticity of funds as there might be in some of our Universities, and it was nobody's business to decide that it would be a good thing to help his and my collaboration to continue. Any additional funds that I could collect from outside were of too uncertain permanence to count on. And so we both have agreed that he ought not to risk losing the chance of a professorship when that came along. There seemed little chance of such in England, and he was not at all anxious to go back into the confusion of Central Europe.

As it turned out, the Neyman-Pearson collaboration wasn't as productive during this period as it had been before E Pearson established the position for Neyman at UCL. A reviewer put it this way:

> It might have been expected, and must have been hoped, that this would lead to more rapid further developments in the Neyman-Pearson approach to the philosophy of statistical procedures. In fact, this did not transpire for several reasons. [E Pearson] had new preoccupations—he now had new administrative responsibilities, and had married in 1934. On his father's death in 1936, he became Managing Editor of *Biometrika*, a post he held, creating a growing reputation for care and thoroughness both for the journal and himself, for the next 30 years. From this time, also, dates his work on successive revision of K Pearson's *Tables for Statisticians and Biometricians*, in collaboration first with L.J. Comrie and later with H.O. Hartley. [E Pearson] was personally concerned in calculating many new tables, notably those of distribution of range in normal samples, percentage points of Pearson curves, and distribution of skewness and kurtosis coefficients. Although Neyman and Pearson produced further joint work, published in the short-lived, but influential, departmentally produced journal *Statistical Research Memoirs* in 1936 and 1938, by 1936 [E Pearson] felt he had gone as far as he wished to go along the path of further elaboration of the Neyman-Pearson approach. In his later work there is, in general, more emphasis on specific applications.[55]

Forty years later, E Pearson himself noted that, by the time of Neyman's tenure at UCL, their "... most exciting work together was finished."[56] Nevertheless, the collaboration with Neyman was identified to be a significant part of E Pearson's four-part legacy, which included the following:

> (i) [E Pearson's] work with Neyman in the development of a fresh way of looking at statistical problems. Although, inevitably, this was associated with considerable controversy, this was not of [E Pearson's] choosing. His attitude is well summed up by a quotation, "The titanic battles which have from time to time been waged across the statistical field were perhaps enlivening to the onlookers, but they were very real and I think harmfully moving to the participants."; (ii) His work in spreading awareness of the potentialities of statistical methods over a wide range of human activity—especially in industry—where they had previously been inadequately (or not at all) utilized; (iii) His editorship of *Biometrika*—an independent journal which continues to enjoy world-wide respect, and has a unique character among statistical journals; and (iv) His work as a teacher and administrator, producing generations of statisticians whose preparation should (and quite often did) produce a well-balanced view of the relation of their specialized training to their functions in society as a whole.[57]

---

[54] Pearson, E.: attachment to letter to the University College London provost, 21 February 1938.

[55] Note: typed manuscript about E Pearson labeled "N. L. Johnson in 'Frequency Curves'," nine pages.

[56] Pearson, E.: handwritten letter to Neyman, J. 12 April 1974.

[57] Note: typed manuscript about E Pearson labeled "N.L. Johnson in 'Frequency Curves'," nine pages.

As we will see later in our story, David was influential in documenting the foundations of statistics, and preparing these generations of statisticians, by taking up the writing of books.

In 1937, David was promoted to assistant lecturer. Looking back, she said: "I'm inclined to think that I was brought in to keep Mr. Neyman quiet" [8]. She taught elementary probability to first year students: "Fundamental theorems in probability with illustrations. Random variates. Theorems on expectation. Tchebicheff's theorem. Binomial and multi-nomial formulae; Poisson's limit. Statement of the Markoff theorem and applications." She also taught, with Neyman, two courses for second year students. One was Sampling human populations. The other was a second course on probability: "Probability laws and their transformations. Theorem of Markoff. Characteristic functions. Markoff-Lyapunov theorem."[58] Co-teaching brought David and Neyman closer together.

In 1938, David was given the PhD in statistics from the faculty of science at UCL. Completion of her PhD took 5 years longer than she intended: She stated on her First Entry Form that her next public examination in view and date was the PhD in 1933, and she repeated that expectation on her Re-Entry form for the next year.[59] Her thesis was titled "Certain Theorems in Correlation."[60] Neyman had been her catalyst. He reportedly encouraged David to take four of her recent publications and submit them as her PhD dissertation [9].

But now David's second mentor was gone. To Berkeley.

Neyman passed away in 1981. David outlived him by a dozen years. He was scheduled to speak in Poland at the European Meeting of Statisticians. A memorial session was held instead. David gave a very personal note: "I knew him over 50 years. He could be quite impossible and we quarrelled strongly every six months or so. But I loved him" [10].

## 4.3 A Calculating Woman

We know from several of David's surviving letters to her supervisor that some of her work as a research assistant included computing results from alternative formulas and comparing the results. She was doing a lot with quadrature formulas. She also assisted with the sale of books and reprints. She was trusted to handle money, invoices, and her professor's bank book.[61,62,63,64]

In 1935, David and Miss Joyce Townend were roommates at 47 Addison House in London. Townend was the research assistant and secretary to the zoology and comparative anatomy department. This department was in David's orbit, because K Pearson moved his office there after he retired. Now David and Townend were jointly purchasing a used car to take on vacation.[65] The car was owned by the National Union of Students. David's possession of it was being held up because the young Austrians were still touring Devon and were not yet back in London.[66] Townend was important to David. We will encounter her again at the end of David's life story.

David's statistical contributions were becoming noted in the statistics literature. In 1931, Student and K Pearson published an article on the z-test in *Biometrika*. In the article, they thanked the 22-year-

[58] University College London Calendar 1937–1938.

[59] David, F.N.: First Entry Form, University College London, 1931–1932; 1932–1933.

[60] Bedford College: archive note, 1938.

[61] David, F.N.: letter to Professor, n.d.

[62] David, F.N.: letter to Professor, 3 July 1935.

[63] David, F.N.: letter to Professor, 29 July 1935.

[64] David, F.N.: letter to Professor, 13 July 1935.

[65] David, F.N.: letter to Professor, 29 July 1935.

[66] David, F.N.: letter to Professor, 13 July 1935.

old David for her kindness in providing "random samples of 10 from a population of Mothers and Daughters with regard to stature. The samples were extracted in the usual manner by aid of Tippett's Random Numbers" [11].

The first author of this article, named "Student," was the great William Sealy Gosset, who is universally known for developing the extensively used t-distribution [12]. Gosset was trained in chemistry and mathematics at Oxford. He was employed by industry, specifically the Guinness Brewery in Dublin, Ireland. During that time, statistics was beginning to be seriously used in brewing. Gosset spent two terms in K Pearson's lab (1906–1907) to learn more. He used his new knowledge to develop statistical methods for small samples like ones he was working with at the brewery. Guinness had a confidentiality rule that prohibited its employees from publishing. That's why Gosset published under the pen name "Student."

Gosset played an important role in the UCL statistical universe and hence in David's early career. It has been written that "... his [Gosset's] endeavor was always to temper... mathematical reasoning with sane common sense" [13, 14]. David benefitted from a relationship with him over the 6 years before he died, and she honored his memory and work. She described him as being "extraordinary," and she expressed the opinion that "he was really the big influence in statistics... I had enormous respect for him, he was a great man" [8].

E Pearson shared David's respect for Gosset, even giving Gosset credit for stimulating his line of research with Neyman. Remembering Gosset 1 year after his death, E Pearson wrote:

> ... while [Gosset's] main papers are few in number, they bristled with suggestiveness. What R.A.F. owed to him - well only R.A.F. knows, but it must have been a good deal. In my case, I know that it was "student" who, probably without ever realizing it, put me on to the idea that we should decide between tests, having regard to the hypotheses alternative to that tested. Whence came all Neyman's and my joint work. To the last "student" was full of these simple but none the less extraordinarily far-reaching ideas.[67]

Indeed, Norman Johnson identified Gosset and Neyman as the two biggest influences on E Pearson.[68]

David's first publication happened in 1932 when she was only 23 years old. She worked on a Bessel function[69] problem together with K Pearson and Samuel A. Stouffer (b.1900–d.1960), an American scholar who was visiting UCL to learn more about statistics.[70] The work was published in *Biometrika*. Stouffer explained how the article came about:

> When I happened to suggest a problem one day in which I was interested, and showed him [K Pearson] a partial solution, he later carried it to his class and devoted several lectures to a new algebraic development which he worked out over a week-end.

It is likely that David was in the class where K Pearson presented his solution to Stouffer's problem. Stouffer went on to explain:

> He asked me to share in its publication, and I collected numerical illustrations, calculations $\chi^2$ [sic] to five significant figures on a Brunsviga. We went over the manuscript together at his country place in Surrey and took a long walk together through the countryside, with its view of the Downs... There was no hint of dissatisfaction with my help, but when the paper finally appeared in *Biometrika*, I noted somewhat wryly that he had recalculated all of my $\chi 2$'s to eight significant figures. And he had probably done it himself, on his own Brunsviga [15].

---

[67] Pearson, E.: letter to Walter Shewhart, 11 July 1938.

[68] Note: typed manuscript about E Pearson labeled "N.L. Johnson in 'Frequency Curves'," nine pages.

[69] Note: Friedrich Wilhelm Bessel (b.1784–d.1846) used these functions in astronomy.

[70] Note: Samuel A. Stouffer later became an eminent sociologist.

Actually, it was David who computed the probability integral tables. K Pearson made David a third author and documented her contribution within a footnote [16]. David was now a scientific author. She had published with Thor himself.

In David's calculations on the K. Bessel function probability integral, she included a quote from Roger Bacon, circa 1260: "He who knows not Mathematics cannot know any other science, and what is more cannot discover his own IGNORANCE or find its proper remedies."[71]

To put David's achievement in perspective, the author of this book published a first paper at age 29. It was a single-authored master's thesis in statistics: "A probability mixture model of completed parity," which appeared in the journal *Demography*. It was considered an achievement that there was such an early career publication. David published 6 years of age earlier.

David performed many mathematical calculations for K Pearson. Because she was terrified of "the old man," she was unable to say "no" to him. If he gave her an assignment at the end of the work day, and she had plans for the evening, she would still have it ready for him when he arrived in the morning, even if she didn't have the time to sleep. Also, she was afraid to tell K Pearson if she jammed the Brunsviga, and so she would fix it herself using knitting needles.

Here is an early artifact of David's work:

> April 8th 1933
>
> Dear Professor Pearson,
> I send you herewith the probability integral of p=0.5 n=15. The probability integral is absolutely correct to five figures. I have sent you also it correct to six figures because except in one or two cases about the mode [?] it is correct. I have differenced[72] the column which is correct to five figures and where the differences seemed a little skew I checked the P.I. by quadraturing[73] from the ordinates with a central difference formula using eighth differences and found that my working was correct.
> I drew out the $\beta$'s on the $\beta 1$ $\beta 2$ diagram last time, so that it was possible to read off the number of cases which lie in the areas of the different types. I don't know if this would be of use but unfortunately I have left it in my desk at the laboratory.
> I hope Mrs. Pearson and yourself are enjoying both the country and the fine weather. I have taken a four days holiday while the weather is so good and am intending to work right through Easter. The computing is going well and I hope to have completed all the tables between and including n=10 to 20 by the beginning of next term. Will you please remember me to Mrs. Pearson.
>
> Yours sincerely,
> Florence N. David[74]

The attachment contained 70 calculations carefully written into 2 columns. Note from the letter the amount of initiative and care that David took with her mathematics at this early age of 24, as well as her graciousness to her employer and commitment to her work.

David did her mathematical calculations on a machine called the Brunsviga, introduced to the market in 1892 [17]. In 1936, the department had 30 Brunsviga calculators.[75] The idea was that each student would have access to one of the machines for working on problems. With a growing department, equipment funds were always needed to purchase new machines, and so the old ones

[71] David, F.N.: document titled "The K. Bessel Function Probability Integral," n.d.

[72] Mallows, C.: email communication to Golbeck, A.L. 16 September 2018. Note: According to Mallows, "'to difference' (a column of figures) is a standard verb. Each element is subtracted from the following one. Then the "second difference" is obtained by differencing these first differences, and so on. David would certainly have used one of the standard formulas for numerical integration to get the probability integral from the ordinates and their differences."

[73] Mallows, C.: email communication to Golbeck, A.L. 16 September 2018. Note: According to Mallows, "'Quadraturing' is not a standard term, but obviously refers to using a quadrature formula."

[74] David, F.N.: handwritten letter, 8 April 1933. Note: David was still calling herself by her full first name, a practice she would soon abandon.

[75] Block and Anderson, Ltd.: maintenance contract, 18 February 1936.

were never replaced. About a third of the machines were over 15 years old; they were noisy and needed frequent repairs.[76]

Many models of Brunsviga were produced over the years. It is likely that David used a Model Nova 13, as these were the most recent models purchased by the department around this time.[77] The machines were used by all of the students. As stated in the University College London Calendar, "Except under special conditions, at least nine hours a week should be given to practical work in the Laboratory. Students will be instructed in the set of tables and calculating machines, and will be given illustrative problems to work out, bearing on the subject matter of the lectures."[78]

The machines were so instrumental in the advancement of statistics as a field of inquiry that K Pearson considered the one on his desk to be a prized possession [1]. Other statistics laboratories had such machines as well. For example, Columbia University in New York had them starting in 1924.

By 1937, the UCL statistics department had 33 Brunsviga machines, 26 of which were used by students for instruction in practical statistics and the rest by staff and assistants. These were different models, slowly acquired starting in 1920. The machines required regular maintenance, and it was a struggle for the department to secure adequate funds from the university to maintain all of the machines, much less purchase a few more each year to keep up with the increase in students.[79] The machines greatly facilitated the statistical work, although today they would be seen less as sports cars and more as tractors.

We fortunately studied statistics when high-speed computers were on the scene, although they had recently been introduced. Computer programs were run in "batch mode" where you had to wait—sometimes quite a while—for a printout of your results. If we practiced patience during these wait periods, then David practiced extreme patience. This is because work on a Brunsviga was tedious.

Here is the process for how you would multiply two numbers:

> ...by entering the multiplicand and then turning the crank "multiplier" times (or vice versa). Models like the Brunsviga allowed certain crank-saving techniques. For example, to multiply 567 by 32:
>
> 1. Pull the levers down to 567 (the multiplicand).
> 2. Pull the knob on the large lever handle on the right outward and rotate once.
> 3. Rest the handle in the claws. Center and lower rows now read 567; top row reads 1.
> 4. Rotate handle once again; center row reads 567 and lower reads 1134. Top row reads 2, noting the number of operations (2 x 567 = 1134).
> 5. Slide the bottom knob to the left to shift the middle number (567) one place, thus multiplying it by 10: 5670.
> 6. Rotate the handle three times, thus adding three copies of 5670 to 1134 and obtaining the product, 18144, in the bottom row.

Moves such as these are nicely illustrated on a Brunsviga Nova 13 (circa 1920) in a YouTube video [18]. The Brunsvigas when used for complex scientific calculations were "laborious, time-consuming, and error-prone," but they got the job done [19].

David estimated that she:

> ...turned that hand Brunsviga roughly 2 million times. We used to use Brunsvigas and they would carry tens in one register, but they wouldn't carry tens in another. Before I learned how to manipulate long knitting needles, which was strictly illegal, I was always jamming the damn thing. When you jammed it, you were supposed to go tell the Professor and then he would tell you what he thought of you; it was really rather awful. Many was the time I jammed the machine and had gone home without telling him [8].

---

[76] Pearson, E.: document about calculating machines, n.d. Note: This document appears to be from around 1938.

[77] University College London Department of Statistics: document about calculating machines in use, n.d.

[78] University College London Calendar 1934–1935.

[79] University College London Department of Statistics: document about calculating machines in use, 1937.

David was no goody-two-shoes. Who could blame her?

This kind of computing work was usually done by women, who worked long hours bending over computing machines while receiving little or no professional recognition [20]. Interestingly, Sharon Lohr conjectured that The reason the eminent statistician Gertrude Cox (b.1900–d.1978) was so interested in experimental designs is because she—like David and other women—had done so much uncomfortable computing using such calculating machines, and realized that less or less complicated computing would need to be done under good experimental designs [21].

David made tables of a Bessel function. At the top of the page, she scratched out a quote by memory taken from Shakespeare's Romeo and Juliet: "Here will I set up my everlasting rest, and shake the yoke of inauspicious stars from this world-wearied flesh." Perhaps this was how David felt after calculating so many probabilities on a Brunsviga.

## References


1. Harte, N. and North, J.: The World of UCL 1828-2004, Third Edition. University College London Press, London (2004)
2. Garwood, F.: The probability integral of the correlation coefficient in samples from a normal bivariate population. Biometrika 25(1/2), 71–78 May (1933)
3. Pearson, K.: On a method of determining whether a sample of size n supposed to have been drawn from a parent population having a known probability integral has probably been drawn at random. Biometrika 25(3/4), 379–410 December (1933)
4. Pulman's Weekly News: Beer student's distinction: Doctorate in philosophy. 19 July (1938)
5. Reid, C.: Neyman - From Life Springer-Verlag, New York (1982)
6. Oxford Academic: Biometrika. https://academic.oup.com/biomet (2021). Accessed 14 September 2021
7. Wikipedia: Bertrand Dawson. https://en.wikipedia.org/wiki/Bertrand_Dawson (2023). Accessed 28 March 2023
8. Laird, N.M.: A conversation with F.N. David. Statistical Science 4(3), 235–246 (1989)
9. Deeply Trivial: Great minds in statistics: F.N. David versus the patriarchy. 23 August 2017 http://www.deeplytrivial.com/2017/08/great-minds-in-statistics-fn-david.html (2022). Accessed 20 December 2022
10. Cox, D.R.: Some pioneers of modern statistical theory: a personal reflection. Biometrika 103(4), 747–759 (2016)
11. Student and Pearson, K.: On the "z" test. Biometrika 23(3/4), 407–415 December (1931)
12. Student [William Sealy Gosset]: The probable error of a mean. Biometrika 6(1), 1–25 March (1908)
13. McMullen, L.: "Student" as a man. Biometrika 30(3/4), 205–210 January (1939)
14. Pearson, E.S.: "Student" as statistician. Biometrika 30(3/4), 210–250 January (1939)
15. Stouffer, S.A.: K Pearson-An appreciation of the 100th anniversary of his birth. Journal of the American Statistical Association 53(281), 23–27 March (1958)
16. Pearson, K., Stouffer, S.A., and David, F.N.: Further applications in statistics of the $T_m(x)$ Bessel function. Biometrika XXIV(Parts III and IV), 293-350 November (1932)
17. John Wolff's Web Museum: Brunsviga Calculators. http://www.johnwolff.id.au/calculators/Brunsviga/Brunsviga.htm (2018). Accessed 30 August 2018
18. Valdes, J.: Brunsviga Nova 13 mechanical calculator (circa 1920). https://www.youtube.com/watch?v=dCbr65JZadQ (2023). Accessed 28 March 2023
19. Columbia University: The Brunsviga Nova 13 mechanical calculator. http://www.columbia.edu/cu/computinghistory/brunsviga.html (2023). Accessed 28 March 2023
20. Golbeck, A.L.: Equivalence: Elizabeth L. Scott at Berkeley. Chapman and Hall/CRC Press, Boca Raton, FL (2017)
21. Lohr, S.L.: Statistical frontiers in survey sampling. The American Statistician 58(2), 145-9 May (2004)

# 5 Serving (1939–1946)

*...Churchill decreed that the scientists should work 7 days a week, 10 hours a day, 70 hours a week with NO VACATIONS...I had just wasted 6 years of my life...*

David, F.N.: Communication to Laird, N. (1989) [1] (p. 241)

## 5.1 Going to War

David served her country as a statistician in World War II. Germany invaded Poland on 1 September 1939 and her country, the United Kingdom, entered the war on 3 September 1939. David served from the beginning of the war. Her service was so important to the war effort that The National Archives of the United Kingdom has a story about her war work on their web site. The title speaks volumes: *How Florence Nightingale David saved lives during the Blitz—with statistics* [2].

The time was right for David to be able to make such substantial impact with her work. If she had been born earlier and served during World War I, or served in the period between World War I and World War II, it might not have been so.

The World War I period had extended from 1914 through 1918. David was in her early childhood. Fienberg pointed out that there was "little or no attempt to use statistical ideas" in World War I. He explained that it isn't clear whether this had to do with the lack of perceived need for statistics or the lack of maturity of the statistics discipline [3].

The period between World War I and World War II had extended from 1918 into 1939. At the beginning of this period, David was in her middle childhood; by the end, she was a young adult. In this 20-year period, the modern foundations of the field of statistics were being laid, so that "the practical application of statistics to military and naval problems could realistically be pursued" [3] (p. 26). In the last third of this period, David was finding her way into the field of statistics: She completed her postgraduate studies, was appointed to a position as research assistant to K Pearson, and was promoted to lecturer. She was prepared for the practical application of statistics to war problems.

Barnard and Plackett had this to say about the importance of UCL to the field of statistics across the World War I period and the period between the two World Wars (altogether, 1914–1939):

> The traditional view of statistics was changed largely by the efforts of a few outstanding individuals directed towards the solution of problems with a practical basis.



A. L. Golbeck, *Florence Nightingale David*,
https://doi.org/10.1007/978-3-031-77768-4_5

The authors mentioned David's mentor K Pearson in particular. They continued:

> The Department of Applied Statistics at University College London was formally opened in 1920... Between 1894 and 1930, University College London was the only place in the United Kingdom where statistical theory and its applications were taught to any depth, and graduate students, whether from this country or abroad, who wished to learn more of the techniques developed by the Biometric School, came there year after year... [4]

At the time, David was at UCL with K Pearson and in the thick of all this. She reminisced about it many years later, calling the period *really seminal* for statistics [1].

The World War II period extended from 1939 through 1945. David was in her early to mid-30s. Fienberg used the word *substantial* to describe the development of statistical ideas during World War II. The foundation of the discipline had been firmly enough established that statistics could be applied to problems of the military, which in turn promoted further development of the discipline [3]. The work started slow and quickly accelerated. David was a strong contributor.

Fienberg noted that the most major of the developments in the field of statistics during World War II took place in the United Kingdom and the United States. It is interesting that at the same time, there didn't appear to be much use of probability and statistics by the Germans [3, 4].

Nearly all of the top British leadership recognized, to a greater or lesser degree, that new statistical services and organizations were going to be needed to successfully manage quantitative information and carry out the war. Churchill grasped this general concept very early, even though he was an often indifferent student at school and not gifted at mathematics.

In fact, Winston Churchill was a major force in the use of statistics in military decision-making in the United Kingdom. "Although no scientist himself, Churchill was a consistent champion of technological innovation and the application of science to warfare. In the view of physicist R.V. Jones, Churchill, '[a]lone among politicians, valued science and technology at something approaching their true worth.' He also esteemed scientists..." [5].

Churchill requested in 1939 that a Statistical Branch be formed to provide advice to the government on scientific matters. The next year when Churchill became Prime Minister, "S Branch" as it was called was developed into the Prime Minister's Statistical Section, to be concerned with all aspects of the war. It remained in existence until 1945 as the Prime Minister's personal advisory board.

In addition to S Branch, there was a Central Statistical Office in the United Kingdom which served the ministers and government departments. Fienberg said this: "Britain was unique in being within the war zone, and yet having a continuity of government and social organization to allow for a truly remarkable set of statistical developments" [3].

Barnard and Plackett used the phrase *completely transformed* to describe the statistical scene in the United Kingdom as the result of World War II. They wrote: "Statistical problems arose in many different aspects of the war effort, and statistical advice was widely sought... their numbers [of statisticians] were wholly inadequate to meet the demand [4]."[1] Statisticians were in such short supply and high demand during World War II that the profession of statistics became a "reserved occupation," meaning that statisticians were exempt from being drafted into the military and often were not allowed to enlist.

When World War II was declared in 1939, the British government immediately moved many professionals, including young ones like Lecturer David, out of their regular positions and into temporary civil service. UCL closed down all research work and dispersed its faculty and staff. Architects went to Cambridge, medicals went to Cardiff, engineers went to Swansea, artists/economists/chemists went to Aberystwyth, Medicals/scientists went to Bangor, other medicals went to Sheffield, and others went to Oxford [6] (p. 201).[2]

---

[1] Note: Barnard and Plackett noted that the number of Fellows of the Royal Statistical Society at the time was only 1,000.

[2] Note: UCL experienced more damage in World War II than any other university in Britain.

David, together with the entire UCL statistics department, was moved to the scientific research division of the Ministry of Supply's Ordnance Board in Oxford. This included statistics department head E Pearson, fellow statistics faculty members N.L. Johnson and B.L. Welch, and D.J. Bishop who was affiliated with the department [4].[3] The only faculty member not moved was Elderton who had an honorary title. The Ministry of Supply had just been formed in order to get supplies to all three branches of the armed forces. The Ordnance Board provided independent and authoritative advice on the development, design, and storage of weapons, ammunition, explosives, armor, and the like [7]. The Ministry of Supply was a logical placement for David and the other statisticians, as it included all of the army research units.

According to Barnard and Plackett, many different statistical problems arose in the course of World War II. The statistical advice of David and other members of this group was widely sought. Broadly speaking, the work concerned "the lethal effectiveness of antiaircraft fire." The statisticians developed mathematical models that incorporated "knowledge of positioning errors, fragmentation characteristics of the shell, and target vulnerability." Even though the group as a whole was large, it was not large enough to meet the demand [4].

David's title at the Ministry of Supply was Senior Statistician and Experimental Officer. She was assigned to analyze the results of guns firing, specifically antiaircraft guns against German bombers that traveled up the River Thames [1]. After doing "anything that came along" while K Pearson's assistant, David felt well prepared to do this kind of work.

Neyman—then in California—was overwhelmed by the prospect of a world war, thinking that the world's situation was "appalling" and saying that it gave him "sickly" feelings. Neyman had family and friends in Poland, and then the Polish capital fell. As Reid put it—referring to the spread of the war to Norway, the Low Countries, and France—"disaster followed disaster." Neyman was concerned about his friends across the globe, including his friends in England. He offered David a position at his Berkeley Statistical Laboratory in the United States, but she turned him down, electing instead to stay in the United Kingdom to do "useful statistical work" and help with the fall harvest [8] (p. 168).

But David ended up being bored with the work she was doing at the Ordnance Board and thought it wasn't doing much good [1]. She stayed with the Ministry of Supply for only one year, until 1940. We found no technical reports authored by David while at this ministry.

A. Bradford Hill (b.1897–d.1991), later described as the world's leading medical statistician [9], was, like David, drafted from an academic position (specifically, reader in epidemiology and vital statistics at the London School of Hygiene and Tropical Medicine) to serve as a temporary civil servant in World War II. He did a lot of work on civilian casualties during the war [4].

Around 1940, after the battle of Dunkirk,[4] David transferred from the Ministry of Supply to the Ministry of Home Security which had also been formed at the beginning of the war. This ministry was set up to direct civil defense, especially air raid precautions. David was appointed to the research and experiments department. Her title was senior statistician, and she worked under Bradford Hill.

Things were getting worse in the United Kingdom. In August 1940, there was threat of a channel invasion and an air blitzkrieg. By September 1940, London was experiencing nightly bombings by the Luftwaffe. Statisticians stopped writing about statistical methods for professional journals, and E Pearson had trouble finding enough material to put together an issue of the journal *Biometrika*. The United States was preparing to enter the war [8] (pp. 161, 167–169, 172–174).

[3] Note: Later the group was located in Sidcup and then South Kensington.

[4] Note: The battle of Dunkirk took place on 26 May to 4 June 1940.

David recalled that, at the Ministry of Home Security, she initially worked on questions such as "How many corpses are there?", "What happens when a bomb falls on the road?", and "Where is the safest place to be when a bomb falls?". She went on to explain that good data were hard to get:

> A bomb falls in the road in London. What happens? Well, everything runs down the middle of the street. So, we had sewers cut, the water cut, the electricity cut, the gas cut. Then you have the business of having to have something done about the water pipes because they're probably contaminated by the sewer. And at the end of the day you say to the gas people, "How long did you take over so and so and how much piping did it take to do something or other?" And they'd say, "Oh, well, we don't know. We sent out two men with a cart, and told them to just go on until they are finished." So you never got any reasonable data [1].

In time, the data situation improved when experiments were designed, data were collected, and data were analyzed.

David was able to do more applied statistics within this assignment. She stayed with the Ministry of Home Security for 5 years until the war ended in 1945 and it was disbanded. After her first year working for this ministry, David felt that she was one of the world's foremost experts on the effects of bombing. David wrote 18 technical reports while there.

David's earliest war reports were methods papers with Frank Garwood, a Forest Products Research Laboratory statistician. David knew Garwood from when he was a student at UCL. They did three war reports together. All were completed in 1942 and were theoretical in nature, applying mathematical statistics to basic bombing problems.

The goal of the David-Garwood papers was to accurately predict and assess damage—at first to houses but then to all kinds of structures—caused by high explosive weapons when used in populated areas. At this time, David and other researchers were still using Brunsvigas or similar mechanical machines to perform calculations. Later they would do their work using machines driven by electric motors [4].

In the first paper, David and Garwood's methods assumed that the bombs fell over an area at random. Experience showed that these methods could reasonably be used when bombing was to an area that was not very large.[5] In the second paper, they assumed bombs fell in a way that followed a definite pattern. The methods in this paper, they argued, were applicable when bombing was to a large area (like an entire city) and centered on a definite target (like the city center) or targets.[6] In both papers, the outcome measure was the proportion of houses destroyed or made uninhabitable.

David-Garwood extended and applied the concept of "vulnerable area" that had been introduced in 1938 by F Yates. The single measure also known as the "standardized casualty rate" was expressed in units of area per unit weight of bomb and was used to predict the consequences of bombs exploding in London and other high density populations.

Specifically, David-Garwood extended the concept "to predict the extent of damage, to compare the effectiveness of different kinds of H.E. weapon and to compare the powers of resistance of different kinds of structure...."[7] The models predicted numbers of survivors of the bombing, numbers who died, and damages to buildings and communications (gas, water, electricity, etc.) Because of David-Garwood, vital services were not interrupted when the Germans bombed London (1940 and 1941).

---

[5] David, F.N. and Garwood, F.: Application of mathematical statistics to the assessment of the efficacy of bombing: Paper I. Theoretical assessment of material damage in random bombing. Ministry of Home Security, research and experiments department, civil defense research committee, N. 338 (1942).

[6] David, F.N. and Garwood, F.: Application of mathematical statistics to the assessment of the efficacy of bombing: Paper II. Theoretical assessment of material damage in systematic bombing. Ministry of Home Security, research and experiments department, civil defense research committee, N. 339 (1942).

[7] David, F.N. and Garwood, F.: Application of mathematical statistics to the assessment of the efficacy of bombing: Paper I. Theoretical assessment of material damage in random bombing. Ministry of Home Security, research and experiments department, civil defense research committee, N. 338 (1942).

The third David-Garwood paper focused on high explosive bomb craters in the circumstance when there is a gas attack on a city. Their methods assessed the degree to which these craters might incidentally be contaminated by the gas bombs. The methods employed a mathematical theory of overlapping that David and Garwood developed in their first paper.[8]

Barnard and Plackett pointed out that manned aircraft raids had pretty much ceased by 1942, the year that David and Garwood put out their three papers. Reid remarked that by 1943 "... things were looking up in Africa, Europe, and Asia." Barnard and Plackett pointed out that then came the flying bomb (V1) and rocket (V2) in 1944. They described the sounds of and actions prompted by the V1 and V2: "There were dives for cover when the sound of V1 appeared, whereas V2 arrived silently and allowed no evasive action." Reid summarized that, in spring 1944, "the war had passed into a new and final phase" [4, 8] (pp. 189–190).

## 5.2 Becoming a Bombing Expert

David and other British statisticians were hard at work on many kinds of bombing studies. American statisticians were also working—but independently of the British—on bombing studies. In addition to the effects of bombing, David worked on the effects of land mines. In general, the expertise of David and the other statisticians was desperately needed on problems for which there were no existing solutions [1].

David had been with the Ministry of Home Security for about 2 years when she produced three technical reports. The first was a study for the National Defense Research Committee of how the positive blast impulse of high explosive bombs varied with distance. She compared two methods when used with encapsulated explosives and provided recommendations about which method to use and under what conditions. This study used existing experimental data.[9]

A second report with the ministry in 1943 was for the research and experiments department and showed that a normal bivariate surface can serve as a reasonably accurate model for observed overall distribution of bombs (by number and by weight) dropped during night attacks.[10] David explained:

> There was a map in London on which was plotted where the [V] bombs had fallen and the problem was to figure out where they were coming from... we found they were changing the silos that they fired them from. Probably they changed about once a week. You see you only had four minutes I think from the time you got the warning to when it arrived, or something like that. And you would have to send out your patrol, and as soon as they started firing someone would drop something on them, so that if they knew where it was coming from they would stand a better chance [1].

The other 1943 report, also for the research and experiments department, was an analysis of a bombing experiment conducted at the officially secret experimental site of Millersford. The effects of a blast were noted when a certain aircraft bomb was set off against certain German Tellermines under certain conditions. Conclusions were drawn about which conditions were advantageous.[11]

---

[8] David, F.N. and Garwood, F.: Application of mathematical statistics to the assessment of the efficacy of bombing: Paper III. The contamination of H.E. craters by gas bombs. Ministry of Home Security, research and experiments department, civil defense research committee, N. 340 (1942).

[9] David, F.N.: The positive blast impulse with distance. Ministry of Home Security, national defense research committee. N. 380 (1943).

[10] David, F.N. (assisted by Latimer, E. L.): The fitting of the normal bivariate surface to observed bomb scatters. Ministry of Home Security, research and experiments department, N. 398 (1943).

[11] David, F.N.: Aircraft bombs against minefields: Millersford Trial, Ministry of Home Security, research and experiments department, N. 275A (1943).

In 1944, David produced a dozen technical reports, one each month on average. There were four in January alone.

The first of the four January papers considered the dispersion of bombs within a stick. The accuracy of bomb placement in a cluster is crucial when trying to clear a path through a minefield for tracked vehicles. The study emphasized the importance of bombs being close enough to each other to detonate anti-tank mines efficiently.[12] A second paper applied static pressure theory to predict detonation distances for Tellermines, validated by experiments. It found discrepancies in seashore sand, leading to further investigation into soil factors and testing of additional variables.[13] In a third study, David utilized bombing density and crater width data from Bologne beach craters to propose an approximate method for calculating the probability of tanks getting stuck near craters, informing recommendations for limiting bombing density to soften beach defenses.[14] The fourth January study examined the ratio of negative to positive blast impulse at various distances from a 1000 lb medium capacity bomb, suggesting a simple approximate rule to predict negative impulse based on positive impulse.[15,16]

There were four more technical reports authored by David in March and April of 1944. One involved experimental sticks of 14 bombs and showed how conclusions can be made from analyses of the stick plots.[17] A second proposed using a 500 lb medium capacity aircraft bomb with specific fill and fuse characteristics to effectively neutralize antipersonnel mines within certain radii from the crater center, reducing risks for personnel crossing minefields.[18] A third was a note that attempted to consolidate experimental evidence on the potential benefits or drawbacks of elevating a line charge for breaching mine belts by Tellermines' blast detonation, tentatively concluding that while raising the line charge could increase path clearance, practical challenges may limit its effectiveness, with no clear advantage observed in overcoming potential shielding of buried mines from blast waves by ground irregularities.[19] The fourth studied mortar bomb effectiveness against S-mines in minefields, suggesting strategies for path clearance and optimal aiming.[20,21]

There were three more reports by David in 1944. One presented results of a trial conducted at the armaments research department, Millersford, on an intermittently filled line charge. It concluded that dividing the line charge had a significant negative impact on its effectiveness against TMP 42 minefields, rendering it unsuitable as an offensive weapon in its tested configuration.[22] In a second study, blast tests on the L.P.Z. (German light anti-tank mine) showed it to be insensitive to blast when buried shallowly and minimally responsive when buried flush, suggesting limited risk of sympathetic

[12] David, F.N.: The dispersion of bombs within a stick. Ministry of Home Security, research and experiments department, N. 313 (1944).

[13] David, F.N.: The blast detonation of Tellermines by the spigot "Hedgehog" projectile. Ministry of Home Security, research and experiments department, N. 321 (1944).

[14] David, F.N.: Bologne craters. Ministry of Home Security, research and experiments department, N. 325 (1944).

[15] David, F.N.: The negative blast impulse of the 1000 lb M.C. bomb, Ministry of Home Security, civil defense research committee, N. 407 (1944).

[16] Note: The summaries in this paragraph were aided by the use of ChatGPT.

[17] David, F.N.: Analysis of experimental stick of 14 bombs. Ministry of Home Security, research and experiments department, N. 323 (1944).

[18] David, F.N.: The use of delay-fuzed bombs for breaching antipersonnel minefields. Ministry of Home Security, research and experiments department, N. 346 (1944).

[19] David, F.N.: The effect of raising a line charge. Ministry of Home Security, research and experiments department, N. 348 (1944).

[20] David, F.N.: Mortar bombs against antipersonnel minefields. Ministry of Home Security, research and experiments department, N. 357 (1944).

[21] Note: The summaries in this paragraph were aided by the use of ChatGPT.

[22] David, F.N.: A comparison of the relative efficiencies of aircraft bombs and line charges in the clearance of anti-tank minefields. Ministry of Home Security, research and experiments department, N. 363 (1944).

detonation if mines are buried at specific intervals.[23] In a third study, countermining trials with the Flying Dustbin against buried TMi 42 mines found it inefficient, requiring prohibitive ammunition for deep minefields, with no statistical difference between firing methods.[24,25]

These reports clearly indicate that David had become a bombing expert. Neyman and other high-powered scientists in the United States who had developed bombing expertise were sent by the United States scientific war strategy leadership to England in 1944 to get additional training and cooperate with experts like David. Neyman would work together with a group at Princes Risborough—David was located at the Forest Products Research Laboratories at Princes Risborough—and another group in London. Princes Risborough had gained a reputation as the "brain of the Royal Air Force," and David was part of that brain. According to Reid, "It was the assignment of Neyman and the other members of the American team to study the problems of target-weapons analysis." David was assigned to mentor Neyman on "the methods of treating such problems as they had been developed at Princes Risborough."

We did not find any reports authored by David in the summer of 1944, likely because she was busy with Neyman and his group. David was living alongside UCL colleagues Phonetics Lecturer Elieen Evans, Italian Lecturer Elizabeth Bigg-Wither, and Joyce Townend in Bledlow, Buckinghamshire [2]. As Reid points out, "All of Neyman's admiration for the English character revived when he visited Miss David and the . . . other 'ladies' from the old laboratory with whom she lived and observed the difficult conditions under which they were living and working without complaint" [8] (pp. 193–195).

The war in Europe ended in June 1945 (it was moving to the Pacific and Japan), and so did David's technical bombing reports. Clearly an admirer of Bradford Hill's, David at the end of the war put together a monograph that described his work during the war with casualty data that resulted from air raids. This monograph also contained what she called a definitive bibliography.[26] It was her last technical war report.

David's war reports were on many very specific bombing strategies. They were meticulously detailed with mathematics, data applications, and figures—just as one would expect from someone like David who produced such a meticulous volume on correlation coefficients. After the three joint papers with Garwood, David was a single author on all of the papers. There is evidence, by April 1944, that David was formulating at least some of her own war research questions. Like many scientists, she had focused on war problems during the war period (i.e., had focused on helping to win the war), at the expense of more normal scientific work on nonwar problems.

Barton summed up David's work for the Ministry of Home Security in this way:

> Her [David's] substantial monographs on mass casualty data and on antitank mines reflect the directly military research that she engaged in. She used to claim some credit for the use of random number tables by British troops laying mines in the western desert (by contrast with the Teutonic practice of laying mines in a rigidly regular pattern which enabled the location of all, once the first few had been detonated) [10].

Barnard and Plackett summed up the statistical war effort in the United Kingdom with these words:

> Peace finally returned, and the statistical scene in the United Kingdom had been completely transformed by the changes which had come in response to the immense demand of war. No other method would have produced these changes in only six years, such is the stability - some would say conservatism - of all British

[23] David, F.N.: Report on blast tests on the L.P.Z. (German Light Anti-Tank Mine). Ministry of Home Security, research and experiments department, N. 419 (1944).

[24] David, F.N.: The use of the flying dustbin against German minefields. Ministry of Home Security, research and experiments department, N. 467 (1944).

[25] Note: The summaries in this paragraph were aided by the use of ChatGPT.

[26] David, F.N. (assisted by Arthur, J. B.): The statistical analysis of mass casualty data. Ministry of Home Security, research and experiments department, N. 544 (1945).

> institutions...the importance of the subject [of statistics] was recognized by the formation of a Government Statistical Service. There was a large increase in the number of people who knew that statistics was an interesting subject, and they had been given an excellent training free of charge. Many of them decided to continue working in statistics. The provision of statistical advice for industry and agriculture had slowly grown between the wars and now expanded more rapidly. Those in charge of higher education realized that an important field had in the past been largely neglected, and during the next twenty years the situation was remedied [4].

David served full-time as a senior statistician, experiment officer, or scientific advisor throughout the entire period of the war and afterward as a part-time consultant while resuming her career as a faculty member at UCL. Her primary statistics war research partners in the United Kingdom were noted men A. Bradford Hill and Frank Garwood, and across the Atlantic, there was Jerzy Neyman at UCB [8].

While David was in the thick of the developments in the United Kingdom, and contributed many ideas, her contributions are not elaborated in either Fienberg's or Barnard and Plackett's accounts, although the latter do mention her involvement in the war statistics work. Fienberg acknowledged missing a number of people—perhaps like David—in his account: "I am sure there must have been a number of people, all over the world, who did important statistical work during World War II, but I was unable to embark on anything close to a proper survey outside the United States" [3]. Our information on David can add to Fienberg's account, and we echo Fienberg that more research needs to be done by others.

Perhaps one of the reasons that David's war work has been missing from the war work discussions is because many of her war papers were for many years classified as Secret. Secret is defined to be "Very sensitive information that requires enhanced protective controls, including the use of secure networks on secured dedicated physical infrastructure and appropriately defined and implemented boundary security controls, suitable to defend against highly capable and determined threat actors, whereby a compromise could threaten life (an individual or group), seriously damage the UK's security and/or international relations, its financial security/stability or impede its ability to investigate serious and organised crime" [11].

The Ministry of Home Security, research and experiments department produced over 500 war technical reports. David was a strong contributor to this collection and to the war effort in general. But she considered her work in the war to be wasted years [1].

Now we return to the article on The National Archives web site, *How Florence Nightingale David saved lives during the Blitz - with statistics*. It concluded that "She may not have thought much of her wartime work, but David certainly made a big contribution to the understanding of the effects of bombing on British cities, and her work was much respected" [2]. We cannot agree more.

## 5.3 Returning to Academic Duties

The UCL campus had been hit twice by bombing during the war. By 1943, the College grounds were largely deserted, with buildings in ruins. Between 1943 and 1945, the College remained deserted, with infrastructure continuing to decay.

Around March 1945, David was still working for the Ministry of Home Security, research and experiments department, Forest Products Research Laboratory, Princes Risborough, Aylesbury, Bucks. E Pearson wrote a letter asking David to return to the statistics department at UCL which was scheduled to reopen.

David received this news with enthusiasm:

> Thank you for your letter which gives me great pleasure. There is a certain amount of tidying up to do here and so I will do as you suggest and let you know at a later date what I think will be the appropriate time for resigning from here. I shall be glad to do anything which you like to suggest towards getting the Department going...[27]

The war in Europe ended on V.E. Day[28] in May 1945. In due time that same year, David requested discharge from her military duties.

The Ministry of Home Security wrote a brief letter to David a month after the official end of the war. It informed her that her last day of service to the research experiments department would be 30 June 1945. She was asked to surrender any official documents still in her possession, as well as continue to maintain confidentiality and "...the same high standard of trustworthiness and discretion..." that she had shown during her time with the department.[29]

David was 39 years old. The first half of her life was over. But she had already made many friends on both sides of the Atlantic because of her single-minded commitment to statistics and research and the life of learning. One could say that the World War II had been won by Oxford, Cambridge, and University of London academics—including David and other master strategic thinkers—as much as by the military [12].

David summarized her war service in four words: "I've wasted six years" [1]. She was a Fabian Socialist in the tradition of George Orwell and an inheritor of the social pioneering instincts of the original Florence Nightingale, after all. In her personal view, the war was mainly foolish damage and death. But she was highly respected for the range and depth of her statistical contributions to the war effort in Europe.

The length of David's war service was extensive. She had gone into service early in the conflict (1939) and had served throughout the war up to the atomic period. However, when on 3 June 1945 a request came from the US Pentagon for her to travel there to serve on a task force to report on the atomic bomb, she refused. David had "done her bit" for the war, and she thought it marvelous that she missed the atomic part of World War II [1].

David did continue to serve in the capacity of a consultant on land mines. When her request to leave full-time military service was granted and she was arranging to return to the university, she asked E Pearson for permission:

> There is one point that I should raise at this stage. At the next meeting of the Landmines Committee Professor Curtis is going to propose that I am appointed scientific advisor on Landmines. This would mean that the S.A.C. [Scientific Advisory Committee] would make what they call an extramural contract with the college and would give the college money for my expenses and for the time spent on their job. It is proposed that this advisory should go on for some time.
> I cannot, of course, accept such an appointment without asking your permission and telling you what time and work it would involve me in. As far as I can see the work should not be much. I have several experimental people trained now and it will mainly be a question of dealing with layouts for any new weapon which is found, and planning the experiments necessary to fill in empirically the gaps in existing knowledge. I am sure that such an appointment would not interfere with my college work.

David asked E Pearson to let her know if he had any objections.[30]

One of the tasks of the Land Mines Committee was ensuring the British beaches that had been mined in case of German invasion be carefully de-mined. One of these was Norfolk Sands. A personal friend of David's died there when a mine exploded during the de-mining process [1]. This helps explain why she continued as a military consultant after 6 wasted years.

---

[27] David, F.N.: handwritten letter to Pearson, E., 1 page, 22 March 1945.

[28] Note: Victory in Europe Day, also known as V.E. Day, was established on 8 May 1945. It continues to be celebrated.

[29] Ministry of Home Security: letter to David, F.N. 9 June 1945. Note: The letter came from the Establishment Division at their Home Office.

[30] David, F.N.: handwritten letter to Pearson, E., 1 page, 22 March 1945.

The academic year 1945–1946 marked the return to London of five sections of the college (agriculture, fine art, political economy, laws, and medical sciences) "which had hitherto remained in evacuation." It also marked the re-opening of the statistics department, along with three other academic units (librarianship, Scandinavian studies, and the chemical engineering laboratory) "after the suspension or interruption of their activities during the wartime period."

The UCL catalog said this:

> Thus the work of the College is being re-assembled in its entirety and in its own buildings for the first time since the outbreak of War in 1939. Emergency temporary measures have had to be taken to provide the essential, though in some cases inadequate, accommodation required to carry on the teaching and social life of the reunited College pending the restoration and rebuilding of its damaged or destroyed portions.[31]

The college was slowly reconstructed by building stone concrete huts. Four of these were built in the Front Quad. Two were used as the Union and hosted a wide variety of student activities. Additional huts were built on Foster Court and in the bomb sites to the rear of the college. The New Refectory, the Library, and several academic departments were established. With better economic conditions, the main building was finely restored, but the General Library was not reopened until 1951, and the building as a whole was not fully reoccupied until 1954 [6]. In the case of statistics, no courses had been given during the war. UCL wanted David to help with reconstructing her old department so that it would be ready to open on 1 October 1945.

David valued her friendship with Neyman, and she was active in politics. She wrote this to him:

> I hope all is well with you and the family. It was good to see you over here and find that friendships stand the test of years. We have just been in the turmoil of a general election of members of parliament. I suppose one's friends automatically think the same as oneself about these things but we had no heated arguments because we all felt the same. I was presiding officer over the notetaking in this district and could not therefore take an active part in going round talking to people but now it is all over I think there is no harm in saying that I hope with all my heart that Labour has gained power.[32] I expect that they will not have done so, but that the Tories will be returned with so small a majority that we shall have to go through the whole business again before the year is out.
>
> I wonder how I'm going to like being back at college. I missed the freedom of academic life enormously at first but have been so comparatively exalted in position this last year with so much power to put my ideas into practice that I wonder if I won't at first find it rather small here being a junior lecturer. However, I shall settle to it with a good grace even though chances of promotion are slight.
>
> Please remember me in a kindly way to your fellow workers... and give my love to Mrs. Neyman and Micky. Joyce and Eileen have asked to be remembered to you. They were anxious for me to make the trip [related to your invitation] so that I could buy clothes without coupons!

Neyman had wanted David to be elsewhere to help him, but she decided to stay in London and help her department. It would be the "fair" thing to do, she told Neyman.[33]

And so David returned to UCL and the statistics department. There she went back to her academic duties as a lecturer on a 5-year appointment.[34] Her specific duties included teaching the first year course on Elementary probability (Fundamental theorems in probability with illustrations, Random variates, Theorems on expectation, Tchebysheff's theorem, Binomial and multinomial formulae, Poisson's limit, Law of large numbers and applications) and a second course on probability (Probability laws, Characteristic functions, the Markoff-Lyapunov theorem, Inequalities for moments, Composition of distribution functions). She was making an annual salary of 575 GBP (about 12,000 in 2022 USD), which was on par with the other lecturers after adjusting for years of service in grade.[35]

---

[31] University College London Calendar 1945–1946.

[32] Note: Labour was a socialist party.

[33] David, F.N.: handwritten letter to Neyman, J. 13 July 1945.

[34] UCL Secretary: letter to David, F.N. 26 July 1945.

[35] University College London: document re nonprofessional staff (full-time), session 1945–1946 (1 October–30 September).

Understandably, the department operated under some uncertainty in the restart. This was reflected in the following statement in the statistics section of the University College London Calendar: "The following is an outline of the courses which would be given in a normal year. The extent to which this programme will be implemented must depend on the students attending and on the availability of full staff."[36]

The department restart must have gone well, because the caveat did not appear in subsequent calendars. In fact, the department was very much alive. David added a few topics to her Elementary probability course (Estimation of parameters of a linear function, Theorem on least squares for two variates) and to her second year probability course (Generalized theorems on least squares, Application of Markoff method for test of a linear hypothesis, Sampling human populations, Applications of the calculus of finite differences in probability).[37]

David's duties immediately after the war also included restoring and supervising the library. Everything had been evacuated for war. David had helped to bring back books stored for safety in Welsh salt mines. Some disciplines, she noted sadly, never recovered their complete collections [1]. This loss was strongly felt by David, because books were where her heart had taken root, long ago at Ivington.

Because of the timing of her return to UCL, David fortuitously missed working on the additional events of the firebombing of Japan in 1945, whereas other statisticians who did work on those events found their work used without their consent. In this case, the work of the statisticians was used by some military leaders and caused more civilian casualties among the Japanese than the two atomic bombings combined.

During the transition to peace time, E Pearson, Welch, Johnson, and David worked part time at the university while they "tied up loose ends" related to their war work. There were many more students wanting to be admitted than they could accept. David wrote:

> The college suffered so much in the air raids that when 1500 students and staff are wandering about one begins to be conscious of the physical limitations of space. The staff as a whole are well cared for but the sight of students standing [not legible] in the quad queuing patiently for a half an hour or more for their lunch is a little depressing. The college is I believe fairly well down the list of building priorities, quite rightly I think because houses for people should come first, but it makes the prospects of immediate relief a pipemaker's dream. Professor Levy was up here the other day. He was very upset about the living conditions of students saying that few were getting enough to eat owing to having to pay such large sums to get a roof over their heads. I haven't heard any of our people complain about lack of food but the shortage of living accommodation is such as to be almost funny. My Burmese students tell me that they are sleeping six to a corridor and of course study under those conditions is impossible. The college is doing what it can. Everyone, provided they have patience to wait, gets a mid-day meal ration free and we keep the libraries and workrooms open until 11 pm. Moreover although nobody likes this situation the college is in, nobody grumbles and judging by the conditions obtaining abroad - we have had several moving lectures from university professors at the continental universities - we are very lucky.

She went on to describe the numbers of students that were coming to her department, and she said: "…It is exhilarating to find that people really want to learn about statistics and that there are more well paid jobs going than there are people to fill them." She went on to write:

> The college is woefully short of books of all kinds but particularly mathematical treatises. The whole of the science library and the library of the London Mathematical Society (which was housed in the college) were burned, a total of 100,000 books in all. There is some scheme afloat to ask American libraries to give us any volumes of which they have a surplus. I know if you feel you can usefully support it you will do so but I thought you might like to hear about it…. We are all so happy to be back that I think we would cheerfully put up with anything. The trouble is to know what is best to do for the students.

---

[36] University College London Calendar 1945–1946.

[37] University College London Calendar 1946–1947.

David was eager to visit Neyman and her other friends in America. She offered to come to teach in the summer if she could be paid, noting that she would need the money to pay for her expenses because she would not be allowed to take money out of her country.[38]

Neyman delivered on David's request and offered her 1,300 USD to teach during the summer sessions from 24 June to 13 September for an average of 10 hours per week.[39] Soon she was making arrangements to visit him at Berkeley. She wanted to bring Joyce Townend with her and asked Neyman for his help to get her friend a visa. She thanked Neyman in advance: "I am grateful to you for giving us the opportunity and I will try to work hard and justify being asked."[40]

The young David was quite humble with Neyman. She wrote to him:

> I have thought of your lectures quite a bit during the past few weeks and remembered your teaching with gratitude. I only regret I was so dumb at the time because the more I learn the more I appreciate what I remember of what you said. I have, at long last, written something on elementary probability. What is good in it you will recognize as your own and what is bad can only be attributed to me. I have said this in the preface and if I can find a publisher with your permission I would like to dedicate it to you. In a way I wish you could subject it to [illegible] criticism before publication but if you did it wouldn't be my writing any more I fear because I will fall short of your standards.

She also told him about the 157,000-word book that she had just finished on The Blast Actuation of Anti-tank Mines. She thought it would probably end up on the top secret list, but it might end up in Neyman's hands through some official channels. She talked about how hard it is to write a first book on a subject.

This letter was important, because it documented David's early recognition that her opportunities at UCL were limited. She told Neyman: "I am 36 now and shall try to move when I'm 40, giving myself 4 years to look around. The money isn't important but it's the recognition of work done that one would like." She wanted to visit the United States to see if she would want to pursue a career there, as Neyman had. She wrote: "I should like to come to the U.S. to see you for many reasons. I don't see any prospect of further promotion here in the department - the department is so small - but before trying to get a job in the U.S. I should like to see what it's like - the conditions of work and living." She pointed out that she currently had a part-time job as Scientific Adviser on Mines, and she supposed she could do that full-time. She then added a comment: "... but I like teaching and should be loathe to give it up."[41] Over time, her like for teaching would change to dislike.

Neyman thought David would make a good temporary hire. This is because she had already published a book of "difficult tables," namely, her book on the correlation coefficient. Also, she had already written two more books, one on probability and one on mines.[42] Neyman also thought there would be a good chance of hiring her, because ex-servicemen who wanted to be taught were "flocking to the University." She would be required to teach 10 hours each week, and he would try to get her a teaching assistant.

Neyman thought he could work through the red tape to make David's appointment happen and pledged to make sure that she had enough salary to cover her airfare and living expenses in Berkeley. He acknowledged that she would have to get permission from her university and wrote: "I supposed there will be difficulty with this. However our University is not devoid of power and may help. We

[38] David, F.N.: letter to Neyman, J. 30 November 1945.

[39] Neyman, J.: handwritten note to David, F.N. n.d.

[40] David, F.N.: handwritten letter to Neyman, J. 5 February 1946.

[41] David, F.N.: handwritten letter to Neyman, J. 19 March 1946.

[42] Neyman, J. [presumed]: note [unknown recipient], n.d.

must only know where and how. Get going!" The offer was different from the one Neyman made to David the year before: This time the payee would be the University, whereas last time it was a war project funded by the Federal Government.[43]

David was thrilled to receive the tentative offer. "How very nice of you to offer me the summer job and of course I accept with grateful thanks," she wrote. She set about getting E Pearson's permission. She would try to travel from England to the United States by boat because they allowed more luggage, and she wanted to be able to buy things in the United States and take them back to England. She wanted to travel by train from New York to Berkeley so that she could see the country and have the train experience. David had preferences about what she would teach and expressed them to Neyman, but she also told him she would be pleased to teach whatever he thought would be most useful in filling the gaps in his curriculum. She wouldn't be allowed to take any money out of England, and she wondered to Neyman about how she would be able to deal with a transfer in New York. She worried about getting to Berkeley on time, said she would be prepared to work hard, and asked Neyman if she could stay at his house if he could "bear it."[44,45,46]

The official invitation came in May. David would receive a total payment of 1,300 USD, half for expenses (which would be exempt from tax with appropriate documentation) and half for compensation. Neyman explained how the tax system functioned. David wanted to keep part of her UCL salary, because she thought it would cost her more than the total UCB payment to cover her expenses while in the United States.[47] Neyman was working on how she would have money during her transfer in New York and even arranged for David to meet with statistician Helen Walker of Columbia University's Teachers College while there.[48] David and Walker would become friends and write long letters to each other covering government and university politics, weather, vacationing, and more.[49]

Neyman thought his house was too small for David to stay comfortably, but he would try to put her up at the Women's Faculty Club which is right on the Berkeley campus. She would teach elementary probability and statistics and statistical methods for quality control for engineers. Neyman also expressed interest in having David give some talks on women's issues:

> It occurs to me that it would be extremely desirable if you would agree to give a few talks on such subjects as Work of Women in England during the War. Here I have in mind specifically your own work and that of Joyce and Miss Evans. As you know, I admired your behaviour very much.

Please don't hurry with a "no" answer about the talks on women, he urged.[50] David did hurry, saying she would be willing to do it. She didn't know if she would be very interesting, but she would try.[51]

David's planning went smoothly, and soon she had approval for all of the arrangements including permissions to leave Britain, enter the United States, stay in the United States for 6 months, and travel by boat. The boat passage would be on the famed Saturnia [13], it would take 10 to 12 days, and

---

[43] Neyman, J: letter to David, F.N. 11 April 1946.

[44] David, F.N.: handwritten letter to Neyman, J. 22 April 1946.

[45] David, F.N.: telegram to Neyman, J. n.d.

[46] David, F.N.: handwritten letter to Neyman J. 16 May 1946.

[47] David, F.N.: handwritten letter to Pearson, E. n.d.

[48] Note: Helen M. Walker was an educational researcher and the first woman to be president of the American Statistical Association.

[49] For example, the eight-page letter, Walker, H.: handwritten letter to David, F.N. and Townend, J. n.d.

[50] Neyman, J.: Letter to David, F.N. 1 May 1946.

[51] David, F.N.: handwritten letter to Neyman, J. 16 May 1946.

she would stay in New York for 1 day with Walker.[52] She reported being in Berkeley June through September doing full-time teaching service at UCB.[53]

Since David was coming from overseas, the Women's Faculty Club would give her priority to get a room. This was, they explained, a courtesy which they wished to extend to her.[54] But the food at the Women's Faculty Club could be a problem, Neyman warned:

> Incidentally, if you don't like to eat at the Women's faculty Club while living there, there is no need to. As I mentioned before, the food there is elegantly served but perhaps not quite sufficient—according to Miss Dix, "for a masculine taste." On my own part I would like to add that the insufficiency is from the point of view of masculine American standards but perhaps not so compared to the English.[55]

Today we find the food at the Women's Faculty Club to be healthy and delicious. The Club currently describes their "food philosophy" on their web page: "locally sourced seasonal produce from organic farms when possible; safe harbor certified fish; Mary's free range non-GMO California chicken; artisan breads by Metropolis Baking Company, eggs from cage-free hens, and organic fair trade coffee" [14]. David said she was willing to stay at the Women's Faculty Club if it was the cheapest, but she wouldn't like it if it fit its reputation of "community living" with monstrous regiments of women.[56] Reputation can be far from fact. We choose to stay at the Women's Faculty Club because we find it to be serene and hygge.

An 22 August 1946 six-page handwritten letter from E Pearson to David while she was in Berkeley showed his close relationship with her. He wrote in significant detail about his recent hospital stay, his move to a new house, his hires in the department, his opinions about new hires ("I should feel very hesitant in taking on a person trained in the Fisher-Yates tradition, unless I already know him personally"), and his dilemmas about assigning office space. E Pearson concluded his letter by saying "I hope you are not being entirely destroyed by hard labour, but I know from experience that it would be stiff work. You will want a week at Walton Cottage to recuperate on return...."[57]

David was experiencing the United States. E Pearson was doing his best to stay connected with her from the United Kingdom.

## References

1. Laird, Nan M.: A conversation with F.N. David. Statistical Science 4(3), 235–246 (1989)
2. Butler, William: How Florence Nightingale David saved lives during the Blitz—with statistics. The National Archives 3 March (2023) https://blog.nationalarchives.gov.uk/how-florence-nightingale-david-saved-lives-during-the-blitz-with-statistics/ Accessed 16 June 2024
3. Fienberg, Stephen E.: Statistical developments in World War II: An international perspective. IN Atkinson, Anthony C. and Fienberg, Stephen E., Eds. A Celebration of Statistics: The ISI Centenary Volume, Chapter 2, pages 25–30, Springer-Verlag (1985)
4. Barnard, G.A. and Plackett, R.L.: Statistics in the United Kingdom, 1939-45. IN Atkinson, Anthony C. and Fienberg, Stephen E., Eds. A Celebration of Statistics: The ISI Centenary Volume, Chapter 3, pages 31–55, Springer-Verlag (1985)
5. Ruane, Kevin: The professor and the prime minister, Frederick Lindemann and Winston Churchill. International Churchill Society (2022). https://winstonchurchill.org/publications/finest-hour/finest-hour-195/the-professor-and-the-prime-ministerfrederick-lindemann-and-winston-churchill/ Accessed 17 April 2024.

[52] David, F.N.: handwritten letter to Neyman, J. 9 May 1946.

[53] David, F.N.: form, 9 September 1946.

[54] pMJD: letter to Neyman, J. 8 May 1946.

[55] Neyman, J.: letter to David, F.N. 13 May 1946.

[56] David, F.N.: handwritten letter to Neyman, J. 16 May 1946.

[57] Pearson, E.: handwritten letter to David, F.N. 22 August 1946.

6. Harte, N. and North, J.: The World of UCL 1828-2004, Third Edition. UCL Press, London (2004)
7. Skentlebery, Norman: Arrows to Atom Bombs: A History of the Ordnance Board. London: Ordnance Board (1975)
8. Reid, Constance: Neyman - From Life. Springer (1982)
9. Armitage, Peter: Fisher, Bradford Hill, and randomization. International Journal of Epidemiology 32(6), 929–931, discussion 931-8 (2003)
10. Barton, D.: Obituary: Florence Nightingale David. Journal of the Royal Statistical Society Series A 157(2), 299–301 (1994)
11. Cabinet Officer: Government Security Classifications Policy: Guidance, Updated 17 July 2023. Accessed 17 January 2024 at gov.uk (2024)
12. Oxford Royal Academy: Oxford Summer Schools. 5 things you didn't know about Oxford and the Second World War (2016) https://www.oxford-royale.com/articles/oxford-second-world-war/ (2022). Accessed 3 January 2022
13. MS Saturnia, Wikipedia https://en.wikipedia.org/wiki/MS_Saturnia (2023) Accessed 2 November 2023
14. The Women's Faculty Club: Dining https://www.womensfacultyclub.com/dining/ (2023) Accessed 1 November 2023

# Part IV

# Up the Ladder

# 6 Renewing (1946–1957)

*Why, sometimes I've believed as many as six impossible things before breakfast.*

Carroll, L.: *Alice Through the Looking-Glass* (1871), quoted by David, F.N. in A First Course in Statistics (1971)

## 6.1 Back to Normal

Life after the war was slowly returning to normal.

In 1946, David was elected a Fellow of the Institute of Mathematical Statistics (IMS). Statisticians or probabilists who are deemed to have achieved distinction in their published work may be elected Fellows of their professional associations. Being named an IMS Fellow is a high honor. Unfortunately, the IMS doesn't have records of the citations for any Fellow elected prior to 1999, and so there is no record of the citation for David.[1] But David was clearly rising in academic stature and had much to be cited for.

In 1947, David was appointed to the UCL professorial board and the refectory committee. In 1948, she was made a member of the Royal Statistical Society (RSS). This recognition surely stemmed in part from her statistical war work. David was nominated by E Pearson and Norman Johnson on the basis of their personal knowledge of her work.[2]

Meanwhile in 1948, David traveled by ship to the United States at the behest of her friend and champion Neyman. There she visited Columbia University and Helen Walker, and then she visited UCB and Jerzy Neyman and taught in the summer session at UCB. This was David's second trip to the United States. The first was in 1944 relating to war work. At this visit, David didn't like the United States' puffed up attitude about winning the war. She was happy to return to the United Kingdom [1] (Fig. 6.1).

David was appointed to be a member of the RSS Council from 15 October 1959 through 8 June 1961. She attended a number of the Council meetings,[3] unlike with the Bedford College for Women

[1] Shaman P. and Gustafson, E.: email to Golbeck, A.L. 24 April 2017. Note: At the time this book was written, Paul Sharman was Scientific Legacy Editor and Elyse Gustafson was Executive Director at the Institute of Mathematical Statistics.

[2] Records of the Royal Statistical Society. Note: Norman L. Johnson went on to become a faculty member at the University of North Carolina at Chapel Hill after leaving Britain.

[3] RSS Council minutes, 14 January 1960 through 8 June 1961.



A. L. Golbeck, *Florence Nightingale David*,
https://doi.org/10.1007/978-3-031-77768-4_6

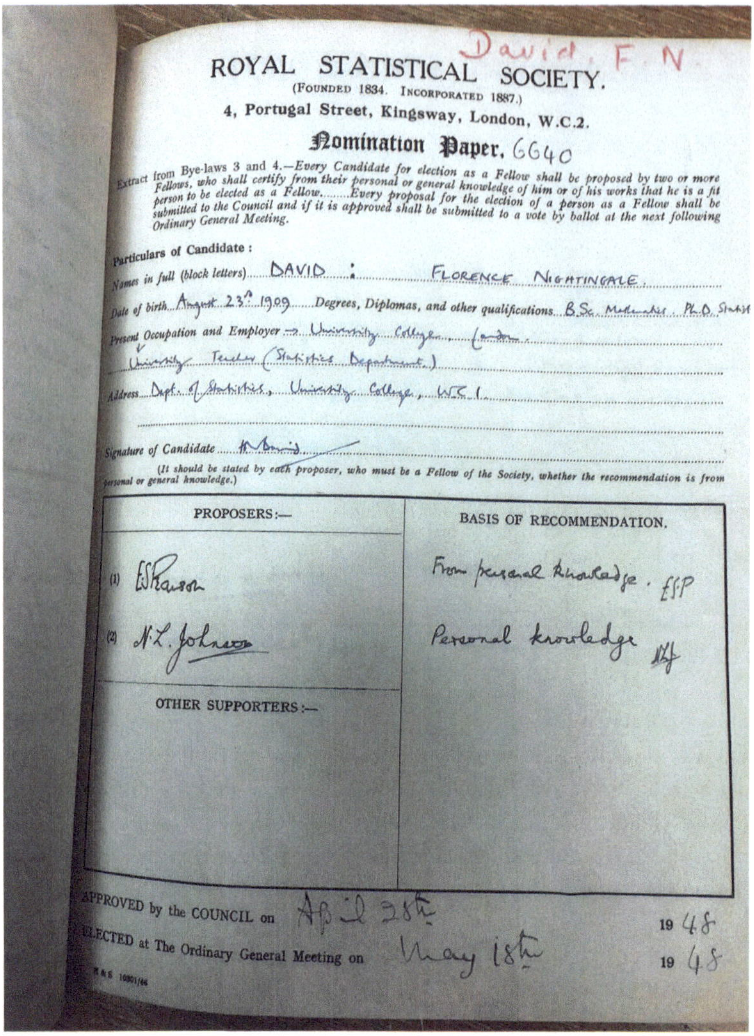
David, F. N.

ROYAL STATISTICAL SOCIETY.

(FOUNDED 1834. INCORPORATED 1887.)

4, Portugal Street, Kingsway, London, W.C.2.

Nomination Paper, 6640

Extract from Bye-laws 3 and 4.—*Every Candidate for election as a Fellow shall be proposed by two or more Fellows, who shall certify from their personal or general knowledge of him or of his works that he is a fit person to be elected as a Fellow.........Every proposal for the election of a person as a Fellow shall be submitted to the Council and if it is approved shall be submitted to a vote by ballot at the next following Ordinary General Meeting.*

Particulars of Candidate :

*Names in full (block letters)* DAVID : FLORENCE NIGHTINGALE.

*Date of birth* August 23rd 1909 *Degrees, Diplomas, and other qualifications* B.Sc. Mathematics. Ph.D. Statistics

*Present Occupation and Employer* → University College, London.

University Teacher (Statistics Department.)

*Address* Dept. of Statistics, University College, W.C.1.

*Signature of Candidate*

*(It should be stated by each proposer, who must be a Fellow of the Society, whether the recommendation is from personal or general knowledge.)*

| PROPOSERS:— | BASIS OF RECOMMENDATION. |
|---|---|
| (1) E.S. Pearson | From personal knowledge. ESP |
| (2) N.L. Johnson | Personal knowledge NLJ |
| OTHER SUPPORTERS:— | |

APPROVED by the COUNCIL on April 28th 1948

ELECTED at The Ordinary General Meeting on May 18th 1948

**Fig. 6.1** Royal Statistical Society Nomination Paper 6640: David, F.N. (1948) [Image credit: Golbeck, A.L.]

Governors where she apparently attended none. David did sometimes speak at the RSS and other professional conferences, but she had as much distaste for conferences as she did for meetings, saying that she "rather hated" them.[4] Being very English, David kept her RSS membership after moving to the United States in the 1960s and even until she died.[5]

At this period in her life, David was living in an apartment at 28 Torrington Square.[6] It was owned by the university. The apartment was 923 square feet on the second floor of the building. There was a living room, two bedrooms, kitchen, bathroom, and W.C. (Fig. 6.2).[7]

[4] David, F.N.: handwritten letter to Pearson, E. 22 June 1976.

[5] Foster, J.: email to Golbeck, A.L. 7 May 2017. Note: At the time this book was written, Janet Foster was the Royal Statistical Society Archivist.

[6] University College London Calendar 1948–1949.

[7] Rent Act 1968: Application for registration of fair rent except in the case of a statutory tenancy which will arise or has arisen by virtue of part I of the Landlord and Tenant Act 1954.

**Fig. 6.2** 28 Torrington Square—David's home while at University College London (2017) [Image credit: Golbeck, A.L.]

Conveniently located on the perimeter of the UCL campus, David had an easy walk from her apartment to her campus office.

David said that, in the 6 years after the war, the students were the most motivated ever: "...they worked like demons." She was teaching at both UCL and UCB, and her observation held on both sides of the Atlantic [1] (p. 241).

Lecturer David continued to develop her lectures around emerging topics in statistics. She wasn't satisfied to develop a course and then teach it the same way for 30 years. In 1947–1948, David added another topic to her second year course on probability (Probability theory in testing statistical hypotheses and in estimation).[8] In 1948–1949, she added two more topics (Derivation of standard errors and Elements of quality control and sequential analysis); she temporarily added a topic to her second course on probability (Stochastic processes); and she taught a special course for research students on Stochastic processes.[9] In 1949–1950, in light of her deepening expertise in probability, David taught two third year courses for research students, one in Combinatory analysis and another in Theory of ranks.[10]

In June 1949, the subcommittee of the professorial board met to consider the claims of David to the title of reader. She had 2 books, 2 war monographs, 33 secret war papers, and 16 original published papers. She had a BSc in mathematics and a PhD in statistics. She had teaching experience at UCL since 1935, with 2 years as a research assistant followed by 3 years as an assistant lecturer and then as a lecturer. She had taught at UCB in the 1946 summer session, and she was a 1946 Elected Fellow of the Institute of Mathematical Statistics. She had worked in the war as an experimental officer, statistician, senior statistician, scientific advisory council member, and scientific advisor.

[8] University College London Calendar 1947–1948.

[9] University College London Calendar 1948–1949.

[10] University College London Calendar 1949–1950.

It was E Pearson who put David forward for the readership. He noted this about her involvement with program development:

> During the last 3-1/2 years the growth in the activities of this Department has been quite exceptional... This development is the result of discussion and experiment in which she has played a most essential part. Many of the ideas which have been incorporated were her's and the scheme for the 3-year B.Sc. Special Degree under the New Regulations now before the Board of Studies in Statistics has been worked out jointly between us.

And E Pearson noted this about David's teaching:

> She is an excellent teacher, who at one end of the scale can present statistical notions in simple form to the non-mathematical biologist and at the other, can give post-graduate lectures on some advanced branch of theory in which she is discovering new results or methods of attack as she goes along. Before the war she had little experience in supervising the work of post-graduates, but in the last two years has shown that she can find problems for others and guide their solution. Her power of handling practical statistical problems is illustrated by the fact that she has been retained as Scientific Adviser to two War Office bodies dealing with mines and mine clearance.

About her research, E Pearson noted: "In addition to the very heavy teaching work of the past 3 years, it will be seen that 7 of her 20 statistical papers belong to this period and she is at the moment full of ideas pointing to further work. One aspect of her work was recently summed up by one of the External Experts of the University: 'What I like about her work is that she doesn't run with herd but sticks to points... which remain to be settled, even if they are not spectacular. We could do with a few more of this type.'"

E Pearson closed his remarks with this note: "The Department of Statistics, under one title or another, has had a long tradition of good workmanship and 'happy' organisation. Dr. David's activities within the Department are such as to continue and strengthen this tradition. She has also played a useful part in the wider field of College activities."[11]

The UCL report that was written for David's promotion said the following about her scholarship:

> Dr. David's scientific studies have lain particularly in the field of the Theory of Probability. Her interest in this direction is important since this subject, which forms the foundation on which the modern theory and methods of statistics must be based, has received in recent years rather little attention in this country. Her original work in the field has both individual character and high quality. In her book which has just been published - Probability theory for Statistical Methods (1949) - she has shown considerable power in bringing together scattered results into a consistent whole and has thus provided a text which should do much to encourage clearer thought in handling statistical problems, both among students and more advanced workers.

And the report said this about David's teaching:

> She is an excellent teacher to all types of students, both in her own and other departments. She supervises post-graduate work successfully and has contributed much to the development of the courses leading to the new Special degree in Statistics, as well as to statistical enquiries in other University departments.

The report concluded by saying: "Her power of handling practical statistical problems is illustrated by the fact that she has been retained as Scientific Advisor to two War Office bodies dealing with mines and mine clearance."[12,13]

With this glowing report, David was promoted to reader. It was "In view of (i) her contribution by research to the advancement of statistics; (ii) her powers as a teacher; (iii) generally her eminence in her subject." With these words, David was also given the status and designation of Appointed

[11] Pearson, E.: handwritten letter to UCL Administration, 23 February 1946.

[12] Gueterbock, E.A.L.: Notice to members, 8 June 1949.

[13] Pearsally, W.H. with edits by Pearson, E.: Report to members, n.d. Note: E Pearson suggested that UCL Professors Haldane and Davenport might also be asked to review and make suggestions to the report.

Teacher.[14] Around the same time, it was arranged that David would continue her service to the Beach Clearance Committee for another calendar year (1950).[15]

## 6.2 Three Books for Teaching

Effective 1 October 1949, David was a UCL reader. She was entitled to hold the title until the standard retirement age of 60. Her starting salary would be 800 GBP a year.[16] Her appointment came in the middle of a university-wide salary review, and David's salary only 1 month later was adjusted to 1,100 GBP.[17] David was so happy about this that she wrote a thank you letter to the Administration.[18]

With the promotion, David's role in teaching expanded. She was assigned an assistant lecturer, Marjorie Thomas, MSc. Thomas assisted David in teaching her regular courses (Elementary probability and the Second course on probability). David also taught a first year Statistics course with E Pearson, and she continued to teach a third year course in Combinatory analysis, co-taught a third year course in Special topics, and taught a miscellaneous course in Elementary statistical methods for biological students.

At the same time that her teaching role expanded, David's scholarship blossomed. She produced three more books over the 15-year period between 1945 and 1959, all related to courses that she taught. These were instrumental in solidifying fundamental probability and statistical ideas, making these ideas accessible to wide audiences beyond the boundaries of her university or homeland, and contributing to what we now view as ubiquitous use of statistical methods in applied research.

David's first book on teaching, *Probability Theory for Statistical Methods,* was published in 1949 [2]. The book sold well, and a second edition was published only 2 years later, in 1951. The book consisted of 16 chapters. David based it on a carefully crafted set of her lectures on probability.

David dedicated the book to Neyman, who she said derived many of the ideas expressed in the book and who she described as her teacher and one-time colleague. He was deeply honored and wrote: "...especially after reading the first few pages and the dedication, I got quite sentimental... The word 'Thanks' is not adequate...."

In the same letter, Neyman talked to David about how he was considering whether to accept invitations to go to England (London and Cambridge) on temporary assignments to teach and give talks. He hoped David would advise him on what clothes to bring for his young son, and he told David he would like to work on some joint publications while there. "Now, if and when I come, let's try to publish a paper or two together - shall we?" he wrote.[19]

David explained that the calculus of probability had not changed much in scope since the work of Laplace in 1812 [3], although it had advanced vastly in terms of mathematical rigor. Her aim in the book *Probability Theory for Statistical Methods* was to "state and prove in elementary mathematical language those propositions and theorems of the calculus of probability which have been found useful for students of elementary statistics... Such theorems as do not appear to have a direct application in statistics have not been considered and an attempt has been made at each and every stage to give practical examples." Accordingly, David moved beyond the usual probabilistic concerns with "...the

[14] n.a.: Recommendation to UCL, 9 June 1949.

[15] Gueterbock, E.A.L: letter to Sir, 20 December 1949.

[16] Statement of Duties and Terms of Appointment, Readership in Statistics Tenable at University College London, 14 June 1949.

[17] UCL Records Department Secretary: letter to David, F.N. 11 November 1949.

[18] Mr. Gueterbock: handwritten letter from David, F.N. 14 November 1949.

[19] Neyman, J: handwritten letter to David, F.N. 18 May 1949.

laying of odds, the tossing of dice or halfpennies, or the placing of persons at a dinner table" to concerns actually encountered in statistical practice [4].

John W. Tukey (Bell Labs and Princeton University) reviewed her book and found little to criticize, noting that David provided "...a clear-headed middle-of-the-road attitude toward the relation of probability theory to the real world" [5]. Cedric A. Smith's (UCL) review pointed out that the book served an important purpose in bringing together "a number of useful results" related to the "simpler parts of probability theory... which otherwise are very much scattered throughout the literature... The style of the book is simple, straightforward, and lucid - refreshingly so; ...the book would be a most useful addition to the library of any statistician or student of statistics" [6]. C.R. Rao (Indian Statistical Institute) commented that "On the whole, the book is well written and the sequence in which the various topics are covered is specially suited to beginners in the subject" [7]. Frank Sandon (schoolmaster in the United Kingdom) noted some unhappy aspects of the book but remarked that "these are all relatively unimportant and can readily be corrected in another edition. Such will probably be found necessary, for there is no other similar collection of the kind available to give sound demonstrations of certain fundamental formulae in the field reviewed" [8].

J. Wolfowitz (Cornell University), referring to David as a "leading British statistician," said that she "writes with an enthusiasm for her subject which is bound to communicate itself to the student. There are many lively and meaningful examples... which should convince the student that the field of application of probability theory goes far beyond problems about gambling or about very dull urns which somehow, in an unexplained manner, always manage to contain balls of several colors. The mathematical argument is kept within the reach of a student with a knowledge of what in the United States is called advanced calculus, or perhaps even less. There is a great lack of good books on probability with these qualities." He noted a flaw in the book, namely, that "no distinction is made between probability theory and the theory of statistical estimation," and he noted some mathematical inadequacies which he said he assumed would be corrected in a next edition of the book [9].

David's second book on teaching, *Elementary Statistical Exercises* [10], was published by her department in 1953. The book was enduring, because 8 years later in 1961, there was a second edition; this time inexplicitly coauthored with E Pearson and published by Cambridge University Press [11]. The book was aimed at instructors of first year practical statistics courses. It provided examples in support of their lectures. It was also aimed at students who wished to learn statistics on their own. Headings and suggestions were included to facilitate the use of the book by such independent students.

The first (1953) edition was developed by David alone. She selected spreadsheets of data used in the UCL first year Elementary statistics course and compiled them into a book. Some of these spreadsheets had been in use for over 25 years in the statistics department. As David explained in the Preface to the 1961 edition, "This selection was reproduced by offset-litho process from typescript. Since this original edition is exhausted a more permanent form of reproduction is now put forward" [11] (Preface).

David was thoughtful about her work. She was usually exemplary at expressing where it fit into the big picture of science. The following justification for her book of exercises is a good example of David's thoughtfulness:

> It is commonplace to remark that to understand a new theory, in no matter what science, it is necessary to practise its application. In mathematics, for example, any fresh theorem is only understood, and its ramifications grasped, if a number of problems all bearing on the theorem are correctly worked out. In statistics where we are concerned with the applications of mathematics to the data of experience this repetition of example and exercise is of special importance.

Thus with the book, David promoted that all students of statistics should have extensive practice in the manipulation of data. Reviewer F. Downton (University of Leicester) pointed out that "Most teachers of statistics have at some time been grateful for the books of numerical exercises compiled at University College, London" [12], and David's 1953 edition was the first of these compilations. David's colleague Johnson, who was still at UCL, put together a set of statistical exercises that could be regarded as a part II to her volume. These exercises covered analysis of variance and associated techniques. They were also published by the department, in 1957, rather than an external publisher.

David's second book on teaching was organized into three parts. In part 1, David organized her practical statistical exercises into 40 chapters. The chapters were different topics, and sometimes she included several exercises on a given topic. Wanting no user of her book to be left behind, David's Chapter 1 was *Use of calculating machines and tables.* In part 2, David included 42 examination questions in practical statistics from the UCL One-year Ancillary Examinations in Statistics and from the BSc General Examinations in Statistics. In part 3, she included answers to the exercises in part 1.

Reviewer John H. Smith (American University) called the book a "workbook" and noted that the "data" were "empirical, synthetic, and self-generated." He suggested that the workbook should be used in addition to, instead of in place of, a good textbook, because although "...the aids are rather good, they are far too brief...." He added that the computations are often "time-consuming" but concluded that "This is a good book of its kind. It has interesting source material not found in typical American workbooks" [13].

Reviewer Bernard Lewis Welch (University of Leeds) remarked that the format of the book was excellent and the collection of exercises was careful and wide ranging [14]. Reviewer Shirley Hitchcock (Open University) agreed that the examples were wide ranging and pointed out that: "The examples are chosen from a wide variety of fields ranging from a large number in anthropology to diverse industrial applications and the inevitable horse kicks in the Prussian army corps." Most of the exercises have to do with sampling distributions and curve fitting.

Hitchcock further pointed out that David presented her numerical solutions to "...a spurious degree of accuracy; e.g. the mean of 83 measurements stated to the nearest integer is given to six decimal places." While the reviewer judged this to be "unfortunate" for the students, we are not surprised given the precision of David's work on the correlation coefficient, the subject of her first book [15].

David's third book on teaching, *A Statistical Primer,* was published in 1953 [16]. It went into a second edition in 1971 under a different title, *A First Course in Statistics* [17]. The first (1953) edition was published as number 31 of Griffin's Statistical Monographs and Courses, Edited by Alan Stewart. This meant it was brief, well written, and reasonably priced. The second (1971) edition had a new format and some additional topics on bivariate data including correlation, bivariate grouped data, and contingency tables. Barton tells us that she wrote the book for first year botanists and that by the time she died, the book was "...still one of the best books on elementary statistics at this level" [18]. As David explained in the Preface to the section edition, the books contained "...a minimum of formulae and the maximum of practical examples...."

In the second (1971) edition, David quoted *Alice Through the Looking-Glass*:

> "I can't believe that!" said Alice. "Can't you?" the Queen said in a pitying tone. "Try again: draw a long breath, and shut your eyes." Alice laughed. "There's no use trying," she said: "one *can't* believe impossible things." "I dare say you haven't had much practice," said the Queen. "When I was your age, I always did it for half-an-hour a day. Why, sometimes I've believed as many as six impossible things before breakfast."

When was the last time you saw a statistics book author conjure up Alice and invoke the reader at the onset to expect the opposite of what is expected? But then, by this time David had moved from the United Kingdom to the United States and herself had passed through the mirror.

David explained the important influence of World War II on uses of statistical methods becoming commonplace. She intended that this book would make it possible for research workers in the sciences to learn and carry out their own basic statistical tests, whereas previously they needed to rely upon a trained statistician. As she explained in the Preface to the first edition:

> The research worker in any of the pure and applied sciences, finds much to wonder at in the ordinary statistician's lack of knowledge concerning the fundamentals of *his* subject; the statistician is always secretly appalled at the average research worker's abysmal ignorance of the elementary algebraic processes. Each is interested mainly in his own subject, and the ideas and techniques of other subjects are of interest to him only in so far as they bear on the central theme of his work. Yet it is out of this cooperation of statistician and research worker that the foundations and facade of modern statistical theory have been built, and it is with this theory that the research worker of the present day reinforces the results of his observations and experimental techniques.

David would have known about such research workers, because she had been lecturing to both postgraduates and undergraduates from across the science departments at UCL.

David added words of caution—or, as she called them, a *protest*—to a Preliminaries section of her 1971 edition:

> Some research workers, possibly with the old tag at the back of their minds that one can prove anything by statistics, carry out their experiments (or collect their data) without really formulating their objective, and they then present the statistician with a mass of figures out of which, they hope, he will extract the truth. It must be said categorically that statistical methods are no substitute for clear thinking, and in fact the use of statistical methods without a great deal of hard and clear thought will only make confusion worse confounded.

Statistics can never replace informed intuition, David explained.

Thus, this book in our opinion has extra historical significance. Pointing out that "The elements of statistical theory are no longer in a state of flux," David seized the moment to document those elements and make them accessible to any research worker in the sciences who wished to *seize the tests*, even workers who knew little or no algebra. David understood that the science researchers could find learning statistics to be tedious, because there is jargon that is unique to the field, but pointed out that it could be equally tedious for the statistician to learn the jargon of the applied field. She also understood, having taught such nonmathematical students (many with at most elementary algebra) for many years, that they were capable of carrying out their own basic statistical tests and designing their own basic experiments without the help of a trained statistician, especially if they were given concrete examples to work out.

David wanted to create an example for each type of science: anatomy, biochemistry, botany, pharmacology, physiology, psychology, medicine, zoology, etc. But she didn't have the time to do them all, so she concentrated on examples that most scientists—namely, anyone "... with a smattering of natural history and a little biology"—would be comfortable with. Many of her examples came from papers published in journals, especially in early volumes of *Biometrika*. David refrained from citing her sources, however, in order to give herself the freedom to "take liberties with the data."

David wrote that she would be satisfied if the book "... acts as a stimulus toward the learning of more statistical methods from other sources." She added that the book was not intended to make scientists into statisticians, writing colorfully that "such persons might with justice say 'O monstrous! but one half-pennyworth of bread to this intolerable deal of sack!'" and pointing them to an introductory statistics textbook. Incidentally, when we read the table of contents, we are impressed by how much it resembles introductory statistics books today, even though this book was written in 1952.

David concluded her Preface by thanking people who helped her. Then she wisely wrote: "Nevertheless, all responsibility ultimately falls on the writer, and it is well perhaps to apologise in advance for the errors of omission and commission which seem inevitably to follow any attempt to put ideas into words."

Reviewer Peter Armitage (University of Oxford) pointed out that there already were many elementary statistics books, but David had found an unoccupied niche: a book, with "many realistic examples and useful diagrams," for scientists. He was very complimentary:

> The special feature of Dr. David's primer is the extreme patience with which the most difficult concepts of elementary statistical method are presented. Scientists meeting for the first time the concepts of sampling distributions or tests of significance usually find them difficult to grasp, and they will welcome the thoroughness with which the arguments here are driven home. Similarly, it is very helpful to find examples illustrated by more than one table, showing the state of affairs at successive stages in the calculations. The corollary of this unhurried approach is that the range of topics covered in the book is rather restricted... there is no treatment of contingency tables, and no mention of bivariate problems. The book is, in fact, accurately described by its title, being intended as an instructor in first principles rather than as a reference book for a wide variety of techniques. As such it is most successful.

Armitage also pointed out that there were "very few typographical slips" [19].

D. R. Wilkie (University of Toronto) wrote:

> The non-mathematician is a timid creature, easily scared by the sudden appearance of an algebraic formula, unable to trust even the clearest mathematical demonstration; but quite able to 'get the idea' if it is suitably presented to him. It is fascinating to watch how carefully the author [David] stalks her prey in the excellently written early chapters of this work, avoiding any sudden movement that might cause him to fly off in alarm. But alas, statistics is really quite a difficult subject, and soon the reader is grappling with a straightforward, and therefore rather difficult, chapter on moments... There is a wealth of numerical illustration to make calculation more certain, and the author takes special pains over a number of points which often give worry to a beginner... It is noticeable that there are no half truths. Exposition is always rigorous with laudable emphasis on basic assumptions...

He concluded by stating that "... this book can provide the complete armoury required for simple statistical work in the laboratory" [20].

Reviewer Frank Sandon (secondary school headmaster in the United Kingdom) wrote that secondary school teachers, in addition to research workers, would find David's book to be useful. He thought that, even though the mathematics in the book were "slight," secondary school students, unless they are in a grammar school, would find the ideas to be too difficult. He noted that most of the misprints are failures "to insert the final bracket in rather lengthy expressions." He summarizes that the book "is nicely got up... a sound elementary treatise... The examples are such as should be of general interest to any keen pupil...." He further recommended the whole volume for school statistical courses [21].

Next came reviews of the 1971 edition. Reviewer J.J. Grimshaw (Beecham Research Laboratory) mentioned a few minor criticisms, one of which was that the statistical tables should not be placed in the chapters where they are first used, but rather at the end of the book where they would be easier to find. But his overall opinion of the book was very positive: "In spite of these minor criticisms, this remains an authoritative text, containing numerous worked examples. The originality of the approach make it a good background text for Part I students and its pronounced biological flavour provides a welcome alternative background to the many other books of this standard with an industrial or economic bias" [22].

Reviewer Alan Sherlock (St. Dunstan's College in Catford, London) thought the book had aged: "This was probably a most successful book in its time. Unfortunately, it still has the appearance and feel of having been published nearly twenty years ago" [23].

Reviewer Stephen L. George (University of Texas) took it from there: "There is no reference to, or mention of, any new development or philosophical shift in statistics during the last twenty years. (The only statisticians mentioned in the text are Karl Pearson and R.A. Fisher)." High-speed computers had come into the practice of statistics. George noted that David neglected to mention these or adapt her book to such major changes in the field. He continued: "Since the original publication of this book, a surfeit of elementary texts have been published aimed explicitly at readers in such fields as engineering, psychology, sociology, business, biology, and medicine... Perhaps the original edition of this text fulfilled a definite need in the early 1950s for an elementary exposition of simple statistical methods. However, such a need is obviously no longer critical and this text must be evaluated relative to the many texts written with similar purposes in mind during the last 15–20 years, many of which are surely more adequate for today's readers." According to George, the field had moved on, and David's original, insightful niche for this book had all but disappeared [24].

At the same time, Barton (then at Queen Mary College) called the 1953 book a "classic textbook." He argued for its continuing utility:

> The first edition... embodies Professor David's teaching of Statistical method to students in the Life Sciences at University College London. Such students were divided into two classes: the more numerate, capable of a two-year course giving a mathematical treatment of Statistics, and the less, into whom over one year the essential ideas had to be instilled without this integument. The book provided a text for the one-year course. Apocryphally, students were allocated according to their answers to the question "What is 1/2 + 1/3?" Those who replied: "1/2 + 1/3 = (1+1)/(2+3) = 2/5" were assigned to the one-year course. The reviewer, who for one mind-bending, if character-building, year had the teaching of this course, can wholeheartedly testify to the merits of the text. The author's judgment is excellent on the presentation of both the key ideas and their corresponding numerical implementation, to science students of little mathematics.
>
> A visit to the bookshop shows how easy it is to write a tedious and boring "cookbook" on the routines of elementary Statistical analysis. It has ever been a puzzle that human writers can bring themselves to this level, but over the years one has been so inured to it that it comes as a pleasant surprise to be reminded that these matters may be treated with freshness and vigour. In style, the book is curiously reminiscent of Fisher's two classics, though David's *obiter dicta* present a humbler and truer picture of the role of Statistics ["It must be said categorically that Statistical methods are no substitute for clear thinking... no really big advance in any subject would have been missed through lack of knowledge of Statistical techniques."] [25]

Reviewer J.S. Maritz (La Trobe University, Australia, and Institute for Biostatistics, South Africa) found the book to have many "excellent" examples and also called it "... a book that all students of statistics can read with profit..." [26].

While putting out the three books on teaching between 1945 and 1957, David also published 34 articles in peer-reviewed journals. She was sole author on 13 of these papers, and she coauthored the rest with N.L. Johnson (11 papers), D.E. Barton (5), M.G. Kendall (4), and P.G. Moore (1). For example, between 1947 and 1950, she wrote about a power function for tests of randomness [27]; a chi-square smooth test for goodness of fit [28]; the probability integral transformation when parameters are estimated from the sample [29]; the application of Fisher's k-statistics [30]; moments of the z and F distributions, building upon herself, E Pearson, Wishart, and others [31]; tables of symmetric functions, referencing Fisher, Kendall, and another [32]; the probability integral transformation when the variable is discontinuous [33]; a two combinatorial test of whether a sample has come from a given population [34]; and other topics. David's work on symmetric functions remained authoritative at the time of her death [18]. Her published papers built upon the work of noted statisticians Bartlett, Fisher, Haldane, Johnson, Neyman, E Pearson, K Pearson, Kendall, Tukey, Wald, Wishart, Wolfowitz, Yule, and others, including herself. It was fitting that she was on the Research Section Committee of the Royal Statistical Society.[20]

## 6.3 Gaining Academic Stature

In 1949 when David became a reader, she became the number two person in the department. There was a single professor (E Pearson), and she was the single reader. When E Pearson became ill for a

[20] Royal Statistical Society Committees List, 1950–1951.

"considerable" length of time, David did both her teaching and his and carried out the "daily routine" of the department without any extra help.[21] Six years later in 1956, David's title became expanded to reader in probability and statistics.

For students in the ancillary course, first year, David together with an advanced student[22] taught an Elementary statistics course. For students in the special course in statistics, first year, David taught the course in Elementary probability; for second year students, she taught the Second course in probability and Systems of frequency curves; and for third year students, she taught Combinatory analysis.[23]

David kept in touch with Neyman across the Atlantic. They exchanged information on students who wanted to study abroad, summer plans, Fulbright and other visitors, numbers of students, and construction updates on David's cottage. She concluded in her letter: "Life pursues its usual monotonous course and I have had enough of ups and downs over the past ten years to be able to savor monotony like a fine old wine. It is a sign of advancing years I suppose to be glad that things don't happen."[24]

Shortly after sending this letter, David received a telegram from Neyman. It offered her a 1-year visiting professor position at Berkeley. He gave her a short deadline of 10 days to accept or decline the offer.[25] David had to turn it down, because the UCL teaching schedules for 1951 were already set.[26] So Neyman asked David to come for spring and summer of 1952.[27] But she again turned him down, saying this: "...we are worse off now than we were before. There is no such thing as being indispensable but if I took leave of absence it would mean very much more work for everyone else and so I am not going to ask for it. Thank you very much indeed for all the trouble which I feel you must have taken. I am very grateful for it and hope that if I ever see my way to coming that I may write to you." David added that she was trying to go to America in the summer of 1952. She wanted to take her mother to visit her sister (and her husband and four children) who had moved to Canada. She was getting help from mathematician Roger Penrose (b.1931—d.unkn.) to line up some paid lectures in Canada. She wondered if Neyman might be able to line up some paid lectures for her in Berkeley, if she promised to "be good."[28]

David told Neyman that she was interested in going to Berkeley in 2 years (1953) and that she could delay any plans for going to Canada until then. The University of California required a loyalty oath at that time (1949–1951), where in order to keep their jobs people would have to pledge that they didn't have radical beliefs. David said that in order to get a visa to the United States she had to disclose such information as "the color of my grandmother's politics," so the oath shouldn't be a problem for her.[29]

In 1951–1952 and 1952–1953, David was scheduled to teach less, which gave her more time for her research: For students in the special course in statistics, first year, she taught the course in Elementary probability; for second year students, she taught the Second course in probability; and for third year students, she taught Combinatory analysis. In 1954–1955 the calendars no longer

---

[21] David, F.N.: letter to Neyman, J. 29 April 1952.

[22] Note: The advanced student was Peter Gerald Moore, who later became a professor in the London Business School and president of the Royal Statistical Society (1989–1991).

[23] University College London Calendar 1950–1951.

[24] David, F.N.: handwritten letter to Neyman, J. 16 March 1951.

[25] Neyman, J.: telegram to David, F.N. 27 March 1951.

[26] David, F.N.: telegram to Neyman, J. 29 March 1951.

[27] Neyman, J.: telegram to David, F.N. 31 March 1951.

[28] David, F.N.: handwritten letter to Neyman, J. n.d.

[29] David, F.N.: handwritten letter to Neyman, J. 15 December 1951.

contained information on who taught ancillary, first year special student, and second year student courses. However, we do know that in 1955–1956, David added a third year course on History of probability and statistics, which she kept the next year when she dropped the Combinatory analysis course. In 1956–1957 she kept the third year course on history and added a third year course in Theory of ranking.[30]

E Pearson made a list of long-term accommodation requirements needed to support instruction in the department. The list included 14 laboratories "each capable of holding 12 large table desks;" 13 offices for regular staff members/assistants/visitors, one of which was for David; a library and rooms for reading, computing, common use, and storage; 2 lecture rooms, 1 dedicated for 30 students and 1 shared for 50; and access to a theater for 150 students. In all, it was 11,000 square feet of working space.[31]

E Pearson's interest in preserving his father's legacy meant preserving the history of statistics at UCL, which also meant preserving the department's history. For example, E Pearson had his father's papers and documents, some of which went back to the 1890s. Over the years, these materials had passed from one unit to another as the organization of UCL changed. When E Pearson became professor and head of statistics at UCL, he reached out to J.B.S. Haldane, who had been appointed professor and head of biometry. E Pearson proposed that the college pay Miss Edmiston, who had accompanied David on her travels for her *Games, Gods and Gambling* book, to catalog the papers and documents. This would essentially create a professional archive that could be split into different departments should the need arise.[32]

David was keeping in touch with her old wartime collaborator, Bradford Hill. In 1950, he was serving as chair, and David was serving as secretary, of the board of studies in statistics. One example of her activity on behalf of the board was to solicit proposals from Haldane for 15 May; and she notified Haldane of a proposal to include demography as an optional subject in the colonial studies curriculum, a proposal which Hill supported.[33] David later described Haldane as "funny" and as having "enormous intuition." She called him a friend [1].

In June 1951, David applied for the DSc degree. She submitted 18 joint papers, organized into the following categories: Correlation, Goodness of fit criteria, Probability integral transformation, Symmetric functions and k-statistics, Analysis of variation, and Probability. She pointed that "All joint papers submitted have arisen from mutual discussion and collaboration. They are truly combined efforts and it is not possible to state further than this."[34] On 2 November 1951, UCL conferred upon David the DSc degree.[35]

A Brigadier General who was director of weapons research congratulated her on her new "distinction."[36] David was still well connected with the research arms of the military.

Around this time, David's appointment as Technical Advisor, Beach Clearance Committee, was ending.[37] David was truly ready to be done with war work. She shared her strong views with Neyman:

> I have been meaning to write to you for some time. I don't like the look of things internationally. One had the feeling that so long as we had a Labor Govt. there would be some slight check on the US. Generals but now

[30] University College London Calendars 1951–1952; 1952–1953; 1953–1954; 1954–1955; 1955–1956; 1956–1957; 1957–1958; 1958–1959; 1959–1960.

[31] Department of Statistics, long-term accommodation requirements, n.d.

[32] Pearson, E.: letter to Haldane, J.B.S. 12 March 1948.

[33] David, F.N.: letter to Haldane, J.B.S. 27 April 1950.

[34] Papers submitted by F.N. David for the examination for DSc Degree (Internal).

[35] UCL Academic Registrar: letter to David, F.N. 2 November 1951.

[36] Hinds, G.H.: letter to David, F.N. 4 December 1951.

[37] UCL Secretary: letter to David, F.N. 13 August 1951.

> we have Churchill. I feel there is no check. I resigned some time ago from all Government work as a protest against the renaming of Germany and Japan and if there is another war I intend to take no part in it. If I refuse to take any part I probably won't be actively presented, since I'm a female, but I expect supplies in the form of salary will be cut off. Is there any chance, in such a situation, that you could find something for me to do in your laboratory? I feel emphatically that I have had my war and I would not be prepared to do anything which would help war projects forward. On the other hand, teaching is my profession and I would be prepared to do this. Think it over and let me know how you feel about it. The college have already asked us what we would do in such a situation and I have replied that I don't intend to do anything which could be described as war work.

David asked Neyman if she could stay in his house when she went to Berkeley, and she offered for him to stay in hers on his next trip to London. She said her department was short on tea, coffee, and sugar and suggested he might send some as a Christmas present.[38]

The board of studies in statistics had jurisdiction over a number of universities. For example, it was responsible for recommending new posts in statistics. For 1952–1957, given the increasing demand for statisticians, they recommended a chair of mathematical statistics in the department of mathematics at Imperial College, a readership in statistics at the London School of Economics, and a readership in economic statistics in the department of political economy at UCL. Under the leadership of Hill and David, the board also at that time recommended the vacant readerships in statistics at Birkbeck College and at UCL be filled and that a central pool of modern computing machines be established, as such machines had proven to be invaluable for statistical research.[39]

The board considered the syllabi for the 1- and 2-year ancillary courses in statistics for botany, zoology, psychology, and anthropology. The idea was that students in these fields in the 1-year courses did not have sufficient mathematical qualifications and that the course should focus on the statistical approach, foundational ideas, and common sense rather than the mathematics. How to treat students in the 2-year courses was more of an open question, and so teachers in those fields should be consulted…"but I think the statisticians' view would be that a student cannot very profitably spend two years on the subject unless he is keen enough to want to grasp some of the mathematics." David, along with E Pearson, prepared the draft syllabi.[40]

Just 3 years after David was promoted to the reader rank, she was again promoted (in 1952) to the upper division of the reader rank. Her teaching and research continued to shine—with seven new journal publications and two books (one new and one second edition)—and E Pearson lauded her contributions to the department's degree programs. Her DSc degree, of course, played a role in securing her promotion, but also there was her election to ordinary membership in the International Statistical Institute, where she was only the third woman ever to be so elected.[41,42] Right before she was notified that she was promoted, she was informed that she was appointed to the University College professorial board.[43,44] David wrote a letter to the administration, saying that she was "very pleased" with her promotion.[45]

David found one of K Pearson's quill pens in a drawer, and she used it to write to Neyman. She had a lot of gratitude for Neyman. She told him: "There are so many kind things to thank you for that I hardly know where to begin." She thanked him for sending a copy of the Berkeley Symposium II

[38] David, F.N.: handwritten letter to Neyman, J. 15 December 1951.

[39] University College London Board of Studies in Statistics: quinquennial policy, 1952–1957.

[40] University College London: packet, 13 pages, n.d. Note: This was presumably to the University College London Board of Studies in Statistics (1951).

[41] Dr. F.N. David, n.d.

[42] Goudswaard, G.: letter to David, F.N. 15 January 1952.

[43] UCL Secretary: letter to David, F.N. 1 August 1952.

[44] UCL Secretary: letter to David, F.N. 6 August 1952.

[45] David, F.N.: handwritten letter to UCL Administration, 7 August 1952.

volume.[46] She also thanked him for getting her elected to the International Statistical Institute, saying "... while it is not a matter of great importance it is gratifying to be elected - if only as a recognition that one is known to a number of people." David then told Neyman about how heavy her workload was. She was picking up a course from Hodges, all of E Pearson's work because he was out for an extended period with a slipped disk, and filling in for Norman Johnson who was going to North Carolina as a visiting professor. David expressed worry that Johnson would stay in the United States. She told Neyman, "I'm not moaning but I hope you will believe I have been damned busy."[47,48] David wondered if E Pearson would be well enough to resume his duties by the start of the next academic session.[49]

On a yet more personal note, David thanked Neyman "very much indeed" for sending her and the department some parcels of tea and sugar. These supplies meant a lot, and she repeated: "It is very kind indeed of you and I say again thank you very much." David noted that she was getting to the countryside every weekend, but she had to work all the time she was there. At least it was cheerful, she said. She then commented on politics: " ... we've been having a to-do because King George died. I was sorry because there goes the last check on Churchill. I can't say I was edified by the mass hysteria that there seems to have been going on. But maybe that is just people's normal behaviour."[50]

There were lively correspondences between David and E Pearson about the history of confidence intervals. He recalled Neyman's early positions on aspects of the topic and cautioned her: "I don't expect it matters what you say at a University lecture, to mathematicians, but it is important to be careful when you write or teach statisticians."[51]

David had an interest in publishing K Pearson's lectures. E Pearson talked about how he himself would have to approach his father's writings in a way that would be objective, but commented that he didn't have time to do it at present. He did think that someone should at least bind the lectures and make them available in the UCL library. He cautioned David that if she published, he and she should be ready to receive criticism for doing it, saying: "... I imagine that you could make quite clear the extent and limits of what you owed to K.P. in your introduction. At any rate, it would be all against the spirit of K.P.'s teaching if I were to prevent you publishing until I had time to do when I retired, what he never got to when he retired!"[52]

R.A. Fisher was a formidable presence in the profession. David was highly critical of him and his so-called fiducial intervals, and so she thought it would be best to wait to publish her writings about his work until after he died. She noted that, when speaking about him professionally, she kept her audiences "chuckling considerably by reading out passages in which Fisher abused other people and then showed how the same criticisms applied to him."[53]

In 1952, David was still describing her work on the history of probability as a hobby. She thought the Royal Society would have some new information for her from the period of around 1690–1720, which she felt was when the "real mathematical foundations of probability took place" via James Bernoulli, Pierre Remond de Montmort, Abraham de Moivre, and perhaps some others. She had their printed work, but now she wanted to access correspondence and other records like meeting minutes.

---

[46] Note: Neyman took the lead in organizing six Berkeley Symposia on Mathematical Statistics and Probability between 1945 and 1972.

[47] David, F.N.: handwritten letter to Neyman, J. 19 February 1952.

[48] David, F.N.: handwritten letter to Neyman, J. 29 April 1952.

[49] David, F.N.: handwritten letter to Neyman, J. 12 May 1952.

[50] David, F.N.: handwritten letter to Neyman, J. 19 February 1952.

[51] Pearson, E.: handwritten letter to David, F.N. 17 March 1952.

[52] Pearson, E.: handwritten letter to David, F.N. 22 March 1952.

[53] David, F.N.: handwritten letter to Neyman, J. 29 April 1952.

She thought the Royal Society might have some records that would be useful to her project. She wrote to them: "I hope from this perhaps to shed more light on who was thinking about probability theory at that time, with the possibility that I may by this means find a new source of material."[54]

In November 1952, Neyman wrote a breezy letter to David. He was in Bangkok at the Food and Agriculture Organization of the United Nations Regional Office for Asia and the Far East. He wasn't sure he would make it to the International Statistical Institute meeting in Rome to see her (due to finances and passport expiration), and so he encouraged David to come to Berkeley the next summer. He complained about the weather in Bangkok which made it difficult for him to travel around and see the country. He told her he was enjoying the work there planning a survey and teaching people from many different countries, and he told stories about the difficulty of teaching people who didn't share a common language. Neyman told David exactly when he expected to be back at his desk at Berkeley.[55]

E Pearson did appreciate David. After spending an evening with her, he gave her and Joyce his best wishes for Christmas and New Year, and he wrote: "Though I might say it, I do feel very grateful for all the staunch help that you always give. If I sometimes say that I want to knock yours and somebody else's heads together, it isn't because I value the heads at the same worth. I rather fancy it is the display of unjustified annoyance on my part on finding that through my own bad planting my College garden is not as lovely as I should like it to be!"[56]

The relationship between E Pearson and David became fragile during his illness. He wrote: "I have a fear that I may have upset you on the last occasion on which we met. If so I am indeed very sorry. It was a particularly churlish thing to do after all you have been carrying this year. I can only say by way of explanation that at times the long drawn out trouble of my back or leg makes me particularly irritable. But apart from that, I am only too well aware that the reason that we don't get research students lies in me. If the head of a department ceases to produce and stimulate, outsiders won't come and any bitter remarks I made were directed against myself, not aimed at you... you may be sure that [in the future] I am not going to leap into any decision without having you with me - at least if I can help it. So please forgive me and forget my inconsiderate outburst!"[57] David gave a quick response, and E Pearson was glad to have it, saying that it relieved his mind to no end.[58]

In 1952 David was working with several data sets, most likely for one of the practical examinations. One was of fatal road accidents which included many particulars of the accidents: date, day of week, time, who the victim collided with, the speed limit, light conditions, conditions of road, the age, sex and date of death of person killed, type of road user, and weather conditions. Another data set was of medical cards for schoolchildren that included borough, number of infectious illnesses, and particulars about the examinations (date, age, height, and weight).[59]

Here is one part of the BSc special examination (part II) statistics practical examination:

1. The schedules supplied you contain measurements made on school-children in the Boroughs of Croydon and Willesden. The Croydon children were mainly born in 1922 and the Willesden children in 1918. Each child was examined and measured on three occasions and the number of infectious diseases it had had by the time of the third examination is recorded. Age is given to the nearest year and month, while height and weight are measured in inches and pounds respectively.

---

[54] David, F.N.: letter to Professor Brown, 7 December 1956. Note: Professor Brown was the Biological Sciences Secretary of the Royal Society.

[55] Neyman, J.: handwritten letter to David, F.N. 6 November 1952.

[56] Pearson, E.: handwritten letter to David, F.N. 21 December 1952.

[57] Pearson, E.: handwritten letter to David, F.N. 22 August 1953.

[58] Pearson, E.: handwritten letter to David, F.N. 28 August 1953.

[59] University College London Galton Laboratory Records: data, David, F.N. 1952.

2. You should take as an important purpose of your investigation the provision, for each sex, of simple charts or tables whereby, given the height of a child and its age to the nearest half-year, it is possible to determine the limits within which its weight would normally be expected to fall. You may interpret 'normally' as meaning the range within which the weights of 19 out of 20 children may be expected to lie. This information should cover the whole age range from 5 to 13 years so that interpolation between the three age classes of the data will be necessary.
3. You should organize your tabulation, computation and analysis so as to provide answers in addition to points of general interest such as the following:-
   (a) Whether there appear to be any differences between the growth of the children in the two boroughs, and if so how this affects your results under (2).
   (b) What are the most striking differences and similarities between the growth of the two sexes.
   (c) The manner in which the relationship between height and weight changes with age.
   (d) Whether there is any evidence suggesting that growth is affected by the number of infectious illnesses which a child has had.
   (e) What are the advantages of getting records of the same children at the three dates?
4. You should conclude by drafting a short Report, avoiding technical statistical language, suitable for a Medical Officer of Health, indicating how your charts or tables required in paragraph 2 may be used and summarising the main conclusions of interest drawn from your further investigations.
5. In addition to this Report, you should put together some Notes suitable for reference to a statistical colleague, indicating the techniques you have used, the tests you have carried out in order to establish conclusions and the nature of any assumptions you have felt it necessary to make.[60]

David's comments on one student's examinations were as follows: "Rather a pedestrian effort. It could have been made much more attractive and informative (for the [Minister of Health]) if he had drawn a few pictures. What he has done is sound enough apart from arithmetical slips. I would like to have seen some effort to use all the data."[61]

Maurice Kendall of the London School of Economics was greatly interested in, and had high regard for, David's work on the history of probability. He was impressed with the large amount of material she was able to amass that had not been already published and expressed a wish that it be published. He had a few points of disagreement on her writing, and he expressed those to her. Kendall also complemented David on her idea of "expanding the variate in terms of the distribution function," which is an idea he said he envied.[62]

Other people of high repute in science regularly sought David out for statistical expertise. For example, J.B.S. Haldane asked her where, if anywhere, a detailed mathematical exposition had been published.[63] For another example, he asked her if she would check two reprints for errors.[64] Haldane was not always complimentary. In one letter related to the statistics BSc special 1952 section part II, he declared, "I return your paper. I don't like it." The letter included detailed criticisms.[65] Neyman and others forwarded to David ideas for her book on the history of statistics.[66,67]

In 1954 David was appointed to the Managing Committee at UCL. She was nominated by the professorial board and appointed by the vice-chancellor on behalf of the Senate. The appointment was for 1 year, and then she was reappointed for an additional year.[68,69,70] David was especially

[60] University College London: BSc special examination (part II) statistics practical examination, 1953.

[61] David, F.N.: Handwritten comments on student examination 24 June 1953.

[62] Kendall, M.: Letter to David, F.N. 2 November 1953.

[63] Haldane, J.B.S.: letter to David, F.N. 4 July 1952.

[64] Haldane, J.B.S.: letter to David, F.N. 5 March 1954.

[65] Haldane, J.B.S.: letter to David, F.N. 23 March 1954.

[66] Neyman, J.: letter to David, F.N. 27 February 1956.

[67] Kendall, M.: letter to David, F.N. 14 March 1955.

[68] UCL: letter to David, F.N. 10 February 1954.

[69] Gueterbock, E.A.L.: letter to David, F.N. 1 March 1954.

[70] Gueterbock, E.A.L.: letter to David, F.N. 7 March 1955.

pleased, because she believed she was the first woman to be elected to be the nonprofessorial staff representative to this committee.[71]

In June 1956 David was appointed to be a governor of the Bedford College for Women. She was considered because of her great credentials, which included a BSc, PhD, and DSc, and that she had achieved the rank at UCL of reader in probability and statistics. The governors were not a small group, numbering 97 in total, but the charter stipulated there could be as many as 300. Most of them were women. David didn't attend the meetings. It might be that she didn't feel she could have much influence on a group that large. David sent no apologies for her absences from meetings.[72] But she did make sure she stayed in touch with the college as late as 1977.[73]

David handled some of the submissions to *Biometrika*, and she could be encouraging. To one author, for example, she wrote: "I think [one referee's] comments are a little severe and think that if you follow the second referee's advise with regard to Section 2 then this should do. [para] I am sorry that you are still having to work over your paper but don't be discouraged by this. We have all of us in the past written our first paper six or seven times and I assure you the process gets easier as the papers increase. Do have another shot at getting it into acceptable form and maybe this time it will be alright."[74] David could also be encouraging to students: "You have obviously put in a great deal of hard work but in my opinion there is not enough original material in it to warrant submission for a Ph.D. with complete chance of success. What to do? Well obviously some more work." She offered a list of suggestions.[75]

David enjoyed the status of being the only reader in the department for 7 years. This was until Norman Johnson was promoted to reader in statistics in 1957. This could be interpreted as a dent in David's status, through no fault of Johnson's. Even though it was common to have multiple readers in a department, regardless, David had been diminished.

## References


1. Laird, Nan M.: A conversation with F.N. David. Statistical Science 4(3), 235–246 (1989)
2. David, F.N.: Probability Theory for Statistical Methods. Cambridge University Press, 1949
3. Laplace, P.S. de.: Theorie analytique des Probabilites (1812) Wentworth Press, United Kingdom (2016)
4. David, F.N.: Preface to Probability Theory for Statistical Methods. Cambridge University Press Second Edition, United Kingdom (1951)
5. Tukey, J.W.: Review of Probability Theory for Statistical Methods by David, F.N. Journal of the American Statistical Association 44(248), 567–568 December (1949)
6. Smith, C.A.: Review of Probability Theory for Statistical Methods by David, F.N. Biometrika 36(3/4), 469–470 December (1949)
7. Rao, C.R.: Review of Probability Theory for Statistical Methods by David, F.N. Sankhya: The Indian Journal of Statistics 10(1/2), 160–162 March (1950)
8. Sandon, F.: Review of Probability Theory for Statistical Methods by David, F.N. The Mathematical Gazette 34(308), 155 May (1950)
9. Wolfowitz, J.: Review of Probability Theory for Statistical Methods by David, F.N. The American Mathematical Monthly 57(8), 570–571 (1950)
10. David, F.N.: Elementary Statistical Exercises. University College London Department of Statistics, United Kingdom (1953)


[71] David, F.N.: handwritten letter to UCL provost, 12 February 1954.

[72] Bedford College for Women: agendas of annual general meeting of the governors of Bedford College, 21 June 1955; 29 November 1956; 28 November 1957.

[73] David, F.N.: letter to Bedford College, 9 May 1977.

[74] David, F.N.: letter to Haight, 12 April 1957.

[75] David, F.N.: letter to Gronow, C. 11 July 1957.

11. David, F.N. and Pearson, E.S.: Elementary Statistical Exercises. Cambridge University Press, United Kingdom (1961)
12. Downton, F.: Review of Elementary Statistical Exercises by David F.N. and Pearson E.S. The Mathematical Gazette 47(360), 167 May (1963)
13. Smith, J.H.: Review of Elementary Statistical Exercises by David F.N. and Pearson E.S. Journal of the American Statistical Association 57(299), 718–719 September (1962)
14. Welch, B.L.: Review of Elementary Statistical Exercises by David, F.N. and Pearson, E.S. Biometrika 49(1/2), 291 June (1962)
15. Hitchcock, S.: Review of Elementary Statistical Exercises by David, F.N. and Pearson, E.S. Journal of the Royal Statistical Society Series A (General) 125(2), 286–287 (1962)
16. David, F.N.: A Statistical Primer. Charles Griffin and Company Limited, London (1953)
17. David, F.N.: A First Course in Statistics. Charles Griffin and Company Limited, London (1971)
18. Barton, D.E.: Obituary: Florence Nightingale David. Journal of the Royal Statistical Society A 157(2), 299–303 (1994)
19. Armitage, P.: Review of A Statistical Primer by David, F.N. Journal of the Royal Statistical Society, Series A 116(3), 325 (1953)
20. Wilkie, D.R.: Review of A Statistical Primer by David, F.N. Biometrika 41(1/2), 281–282 June (1954)
21. Sandon, F.: Review of A Statistical Primer by David, F.N. The Mathematical Gazette 38(325), 239–240 September (1954)
22. Grimshaw, J.J.: Review of A First Course in Statistics by David, F.N. Journal of the Royal Statistical Society Series D (The Statistician) 21(1), 96–97 March (1972)
23. Sherlock, A.: Review of A First Course in Statistics by David, F.N. The Mathematical Gazette 56(397), 251 October (1972)
24. George, S.L.: Review of A First Course in Statistics by David, F.N. Technometrics 14(4), 983–984 November (1972)
25. Barton, D.E.: Review of A First Course in Statistics by David, F.N. Journal of the Royal Statistical Society Series A(General) 135(4), 612–613 (1972)
26. Martiz, J.S.: Review of A First Course in Statistics by David, F.N. International Statistical Review 41(3), 397 December (1973)
27. David, F.N.: A power function for tests of randomness in a sequence of alternatives. Biometrika 34(3/4), 335–339 December (1947)
28. David, F.N.: A chi-square 'smooth' test for goodness of fit. Biometrika 34(3/4), 299–310 December (1947)
29. David, F.N. and Johnson, N.L.: The probability integral transformation when parameters are estimated from the sample. Biometrika 35(1/2), 182–190 May (1948)
30. David, F.N.: Note on the application of Fisher's k-statistics. Biometrika 36(3/4), 383–393 December (1949)
31. David, F.N.: The moments of the z and F distributions. Biometrika 36(3/4), 394–403 December (1949)
32. David, F.N. and Kendall, M.G.: Tables of symmetric functions—Part I. Biometrika 36(3/4), 431–449 December (1949)
33. David, F.N. and Johnson, N.L.: The probability integral transformation when the variable is discontinuous. Biometrika 37(1/2), 42–49 June (1950)
34. David, F.N.: Two combinatorial test of whether a sample has come from a given population. Biometrika 37(1/2), 97–110 June (1950)

# 7 Transitioning (1957–1967)

*I have heard it suggested that her [David's] appointment to a professorial chair would tend to prejudge the appointment of a successor to yourself [E Pearson]. I can think of no better successor.*

Haldane, J.B.S.: letter of recommendation for David, F.N. (1957)

## 7.1 Hitting the Invisible Barrier

E Pearson wrote to J.B.S. Haldane asking for an opinion of David's work and whether she was worthy of a promotion to the rank of professor. Haldane replied with a very supportive letter:

> Dear Pearson,
> You ask me for my opinion on Dr. F.N. David. Some of her work is of a very high quality. I think particularly of her recent work on symmetrical functions, which is a key to a great variety of problems. I am particularly impressed by her overall knowledge of statistics, and her very critical attitude to statistical theories. She has constantly helped me in my own work. During the second world war she was engaged on practical problems. Little if any of this work was published, perhaps because its publication would have discredited some people in high places. However, it is important that she has a very broad practical experience. I cannot personally speak of her powers as a teacher. But her small books, based on lecture courses, are very clear, and useful to applied statisticians like myself. And she is most lucid in her verbal explanations to me.
> I have heard it suggested that her appointment to a professorial chair would tend to prejudge the appointment of a successor to yourself. I can think of no better successor. In particular a number of people with Fellowships of the Royal Society and the like are narrow specialists compared with Dr. David, and would be likely to lower the high standards of teaching and of tabulation which the Department has achieved. She has also a considerable experience of administration, which is very important in a large department.
> For all these reasons I emphatically support her claim to a Professorship in the College. I consider her to be well above the standard of a number of our professors though not, like a dozen or so of them, a person with a great international reputation. Nevertheless it is worth noting that she has refused at least one professorial post in the United States. I cannot think that the College would regret her appointment to a Professorship.
> I shall be glad to make a more formal report on her. If this is desired, please send a complete list of her published work to me at the Indian Statistical Institute. My papers have just been packed up, and I cannot undertake a search in libraries.
>
> Yours Sincerely,
> J.B.S. Haldane.[1]

---

[1] Haldane, J.B.S.: Letter to Pearson, E. 18 July 1957.



A. L. Golbeck, *Florence Nightingale David*,
https://doi.org/10.1007/978-3-031-77768-4_7

This is the kind of letter that most people would be happy to have written about them, especially by a person with Haldane's stature [1].

But E Pearson wasn't happy about the letter. It wasn't what he wanted to hear. He was thinking about the future of his department, and David was such an integral part of it that his thinking included her. UCL had a mandatory retirement age, and he was approaching it. This meant that a successor would need to be found. E Pearson wanted to influence that decision.

The choice of successor would affect E Pearson's legacy. He was sufficiently concerned about this matter that he wrote to his provost about it in November 1957, three full years before he was scheduled to retire. He wanted his provost to have time to think the matter over before they met, and so he wrote the following in the first two paragraphs of the letter:

> There are really two points at issue:
>
> (a) the possibility of conferment of the title of Professor on Dr. F.N. David;
> (b) the time of my own retirement.
>
> As regards (a), the suggestion has been put to me and supported by a number of persons, and there is without doubt a case for consideration. Dr. David's publications are extensive and are continuing without any signs of flagging; some, if not all, of them are of very good quality. She has been responsible for the supervision of the work of more than half of the statistical post-graduates during the past 10 years, and she has been closely concerned with the development and expansion of our degree courses since the war. On the other hand, I am inclined to think that she is not by temperament the right person - or the best person - to become head of what is really the senior Statistics Department in the country.

The Cambridge Dictionary defines *temperament* to be "the part of your character that affects your moods and the way you behave..." [2]

Lehmann was clear about what had happened: "When Egon Pearson retired in 1960, David was the natural person to succeed him. However, prejudice against women prevented her appointment, and instead an outsider, Maurice Bartlett, was brought in to succeed Pearson" [3] (p. 117).

If E Pearson was lukewarm about David's qualifications as a scholar, she felt likewise about him. She told Laird: "[E Pearson] never really did anything very creative, but he had a very good critical brain" [4].

In the third paragraph of the letter, E Pearson continued to write about his own retirement plans. With a 2-year extension, he could remain in his current position until 1962. But his inclination was to leave in 1959 if he could find some part-time work. E Pearson wanted to get the matter settled soon, thus putting into play his plan of making sure David would not succeed him. This would be "for the good of the department," he wrote. His letter continued: "Suitable candidates for the post are very few, but an obvious and happy choice would be M.S. Bartlett of Manchester, who I know would be interested." E Pearson appealed to the provost's sense of competition and claimed that Cambridge might also try to recruit Bartlett to a chair position.

Then in the fourth paragraph of the letter, E Pearson got more specific about his concerns in regard to David:

> Be that as it may, if the conferment of title on Dr. David were to be taken up, it should be done on the merits of her work and with no particular expectation that she would succeed to the headship of the Department. At the moment I would say that Bartlett is the man we should gain most by attracting and he is also probably the one person whom Dr. David would find it easy to work with. There is a very great risk that she will go elsewhere, perhaps to the U.S., were she to be left as a Reader with a new and younger Professor. This would be a great loss to both Department and College.

It appears that David was supposed to be a butterfly pinned to the wall and remain in her current position.

In conclusion to his letter, E Pearson told the provost that he had already talked *in confidence* [our emphasis] with Maurice Bartlett, David Kendall, Bradford Hill, and another person about these matters. E Pearson told the provost: "Their views naturally are not quite identical and there are clearly a number of difficult issues...." E Pearson then asked the provost for a meeting.[2]

E Pearson clearly had Bartlett in mind to succeed him. Also, given the way the letter to the provost read, Bartlett had to be aware of this. Yet E Pearson asked Bartlett his opinion about David.

The provost advised E Pearson that it would be best to discuss the matters further in person rather than in a letter. He suggested they meet in mid-December.[3] The meeting occurred, and the provost documented that no action would be taken at that time. While it was clear that they were still thinking about possible conferments of the professorship title, neither David's name nor the names of others were mentioned in the meeting notes.[4]

David's research interests at this time were symmetric functions, which gave rise to publications and tables that she co-authored with Kendall. She was almost finished with her next book, on combinatorial chance which she hoped to publish with Cambridge University Press. Her theoretical papers had to do with ordered variables, with a combinatorial background. A few of these papers had to do with "quick" tests, another few had to do with ordered points distributed at random along a line that is of fixed length, another had to do with runs of more than two elements, and a few others had to do with models giving alternatives to randomness where there are two alternatives. Although David wrote all of these papers with co-authors, the ideas and drive were hers. She developed many statistical tests. Her work indicated that she was a master at algebraic manipulation.

Some of David's work, in the words of E Pearson, "involved a great deal of mathematical labour and ingenuity" but nevertheless went unpublished. He didn't like David's choice of assumptions to underlie her work and opined that she was more interested in mathematics and probability than in statistics. E Pearson was able to shelve David's work because he was a double gatekeeper, being both her chair and the editor of *Biometrika*. E Pearson admitted that he was quite possibly "unfair" and "damping" when it came to David's work.

Conversely, E Pearson also admitted that David had significant "fertility of ideas" across many fields. He also thought she had stimulated younger faculty members and students of many different capabilities. She was very helpful in the department's teaching program, and she was very helpful to other departments.[5]

Bartlett wrote to E Pearson that he "very much" appreciated "the difficulties you feel over F.N.D., and your further remarks are more helpful...." Bartlett said that, when he met with E Pearson, he came to the conclusion that F.N.D.'s case was borderline, but she "deserved promotion to a chair, bearing in mind her research, assistance with research, teaching and general services to the college." He told E Pearson:

> I found your present summing up of her [David's] work very fair, and I should not differ appreciably from your own opinion on her work. Leaving me out of the picture for the moment, I agree that she would be a strong candidate for your position when you retire. However, my opinion of her research work does not stand high enough for me to feel happy about her being your successor, and I would still prefer one of the younger people who might be eligible (e.g., in London alone there are D.R. Cox and P. Armitage). It is this which makes any recommendation for her promotion so tricky, as I feel it should <u>not</u> carry any guarantee or implication that she would be your successor. I do not know how F.N.D. will react to particular individuals if one of these were appointed, but whoever it was should be able to stand F.N.D. leaving at that stage if she felt inclined. Of course she may be seeking some thing new in what will happen, so that she can plan her own line of action without

---

[2] Pearson, E.: Confidential letter to the provost 29 November 1957, Note: Ifor Evans was the provost at the time.

[3] University College London Provost: Letter to Pearson, E. 21 December 1957.

[4] University College London Provost: Memorandum to the file 8 January 1958.

[5] Pearson, E.: Handwritten letter to Bartlett, M.S., 3 pages n.d.

> delay, but, while this is understandable, it would hardly (if others agreed with my opinion) be possible to predict the position for her precisely.

Thus, Bartlett suggested that he could be left out of the picture for the assessment of David. But then he did an immediate reversal and put himself at the center of the picture, writing:

> My views [given] above seem to me to make my own future less important to what may be recommended, but in so far as you feel it may help you, I will try to indicate how I feel. While I am quite happy at Manchester, I should probably be pleased at an opportunity to come South again (with its greater facility for London meetings, etc.), if one materialized. At the moment, from the point of view of inclination, I would rank your Chair, if I ever have the honour of being elected to it, rather on a par with any at Cambridge, and I certainly do not put it an inevitable second to Cambridge. If to this you add that (a) there is at present no chair at Cambridge and (b) ever if there were, I would not necessarily be invited to fill it (e.g., the electors might look for a stronger mathematician), I think the chances of my being a candidate for the U.C. post are quite high!

Then Bartlett arranged to meet with E Pearson while in London in February. He concluded by telling E Pearson that he did not agree with the assessment of David, where Haldane wrote that "one can think of 'no better successor'" to E Pearson.[6]

We note that this correspondence constitutes a case study in conflict of interest. Concurrently, E Pearson worked to eliminate David and recruit Bartlett. Also concurrently, Bartlett promoted himself while helping eliminate David.

When they last met, which was 3 weeks earlier, E Pearson and the provost agreed to put off for another year the decision of whether to promote David to the rank of professor. But now E Pearson was having additional thoughts. Marking his letter "confidential," he wrote to the provost:

> I was not, on reflection, altogether happy in my mind about this decision, because I was not and am not at all sure that events will mark time. I therefore had a talk with [?] to whom I knew Dr. David had expressed her fears about the future. This talk helped to bring out the alternatives with which we are faced and, as a result I approached M.S. Bartlett again. He is the most important External Expert as well as a very balanced and fair-minded person. On hearing from Bartlett, I have had another talk with [?] and on his advice. I should like, if I may, to have another talk with you. First however you should see Bartlett's letter to me which I enclose to make it more explicable. I send also a copy of some notes which I sent to Bartlett with my own views on F.N.D.'s work. In essentials, his assessment seems to be the same as mine.
>
> The point which [?] saw and on which I should like your advice is this. The essentials of the problem are clear now and will be unlikely to have altered much in a year's time. Might it not be better for the Appointments and Promotions Committee to be made aware of the problem now by my putting forward a proposal for conferment of title? No doubt the A and PC may decide to postpone any decision, but: (i) the members of the committee would have become aware of the problem and would not be taken by surprise if Dr. David were to decide during the year to resign and go elsewhere; (ii) the decision to wait would be an A and PC one, possibly based in part on finance, and we should avoid the personal difficulties which might arise if it seemed that I alone was the stumbling block to the proposal going forward.
>
> My impression is the fact that the proposal was, as it were, on the College agenda, might prevent Dr. David deciding in a hurry to throw in her hand, which is what I am rather afraid of.[7]

At the end of January 1958, E Pearson asked to talk to the provost again. A meeting was agreed upon.[8]

The provost set the meeting for mid-February 1958 and invited two other professors, one in mathematics (Harold Davenport) and the other in chemistry (Christopher Ingold).[9,10] The provost asked for a pre–meeting with Davenport and asked him to read some papers before the meeting.[11] The question of whether David should be given the title of professor was discussed among the four

[6] Bartlett, M.S.: Letter to Pearson, E., 4 pages 18 January 1958.

[7] Pearson, E.: Letter to provost 28 January 1958.

[8] University College London Provost: Letter to Pearson, E. 29 January 1958.

[9] University College London Provost: Letter to Davenport, H., Ingold, C., and Pearson, E. 10 February 1958.

[10] University College London Provost: Letter to Davenport, H. 3 February 1958.

[11] University College London Provost: Letter to Davenport, H. 3 February 1958.

(E Pearson, Davenport, Ingold, and the provost) at the mid-February meeting. They agreed that "it would be unwise to proceed with this proposal." Ingold drew the short straw and was designated to discuss the agreement with David.[12]

In the meantime, the university was offering higher retiring ages to faculty members. David was around age 50. She confirmed that she wished the higher retirement age to be applied to her.[13] She received an extension of her appointment as reader to the year in which she would reach age 62.[14]

## 7.2 A Foot Out the Door

Neyman set his sights on bringing David to Berkeley from the time he moved there from UCL in 1938. Back then he invited her to work in his Statistical Laboratory, and she preferred to stay in England [5] (p. 168).

By mid-March 1958, David had an invitation to lead a series of research seminars at UCB. David made a formal request to the UCL provost for a leave of absence from her UCL position. She proposed that the leave be without pay and extend from 20 June to 2 July 1958.[15,16] Because her request was for more than a week, it needed to go before the college committee. However, the provost told David that, if there was urgency, she should assume the committee would grant her the leave and go ahead with her arrangements.[17]

David received certification and notification that she was granted the leave of absence and would be able to return to her UCL position in June 1958.[18,19] Furthermore, she was given the leave with pay, which had to please her very much.[20] The administration asked E Pearson to confirm that he had made satisfactory arrangements for the continuation of David's work for the statistics department during her absence,[21] and he did (Fig. 7.1).[22]

David wrote a gracious note to the UCL secretary: "Thank you for your letter about my leave of absence. If appropriate, will you please tell the College Committee I am grateful for the privilege."[23]

The next year, David received another invitation to UCB, this time to give both lectures and seminars. David explained to her provost: "I am very pleased to have been asked since a number of Russians are going to be there including the great A.N. Kolmogoroft [sic] of Moscow.[24] Kolmogoroft is the greatest man in the world in the field of mathematical probability and I shall be able to attend his seminars and discussions while I am in Berkeley."[25] E Pearson vouched to the UCL provost that he had made arrangements for her departmental work.[26]

[12] University College London Provost: Note to the file 14 February 1958.

[13] David, F.N.: Note to the University College London administration, n.d. but after 30 January 1958.

[14] University College London Academic Registrar: Letter to David, F.N. 13 January 1958.

[15] David, F.N.: Letter to University College London administrator 16 April 1958.

[16] University College London: Resolution, leave of absence 6 May 1958.

[17] University College London Provost: Letter to David, F.N. 17 April 1958.

[18] University College London Secretary: Letter to David, F.N. 17 April 1958.

[19] University College London Secretary: Letter to David, F.N. 8 May 1958.

[20] University College London Secretary: Handwritten note 22 April n.d.

[21] University College London Secretary: Letter to Pearson, E. 22 April 1958.

[22] Pearson, E.: Letter to University College London secretary 28 April 1958.

[23] David, F.N.: Handwritten letter to University College London secretary 19 May 1958.

[24] Note: This was Andrey Nikolaevich Kolmogorov (b.1903–d.1987).

[25] David, F.N.: Handwritten letter to the University College London provost 6 January 1959.

[26] University College London Provost: Letter to Pearson, E. 7 January 1959.

**Fig. 7.1** David at Berkeley with fellow statisticians J. Neyman and E. Fix (1960). (Image credit: Reprinted with the permission of the Institute of Mathematical Statistics)

The provost told David, "I am very glad that this opportunity comes when there is such an interesting Group in Berkeley. As you know, your studies are very difficult for the layman to follow, but I do know how much your own contributions are appreciated and I am sure that a very warm welcome will await you in California. [para] I do hope that you have a very interesting time."[27] The provost told David to go ahead with her arrangements, and soon thereafter she received formal certification and notification that she was granted the leave of absence with pay for the last 5 weeks of the 1959 summer term.[28]

In 1959–1960, David was one of two readers in the UCL department of statistics. Her readership was in probability and statistics. Norman Johnson's readership was in statistics. The philosophical foundation of the curriculum was expressed in the UCL Calendar: "The courses aim at providing students with the statistical tools required in a wide variety of practical problems, ranging from the biometric and sociological to the engineering and industrial. While the approach is a mathematical one closely associated with the theory of probability, special emphasis is laid on the practical methods of application." The instructors of record were noted in the UCL Calendar for the Third Year Courses, with David teaching two courses: Theory of ranking and History of probability and statistics.[29]

David had earned a positive reputation with the US military. They asked her to fly to the United States 4 days before she was scheduled to be in Berkeley.[30] The provost granted her permission to begin her leave of absence early in order to consult for the military.[31,32] Some of her work for the military resulted in white papers. The first was titled An "effective" stochastic model for interpreting the correlation between the precipitation at two stations. She labeled this *Scientific Report No. 1* and

---

[27] University College London Provost: Letter to David, F.N. 7 January 1959.

[28] University College London: Resolution, leave of absence 3 February 1959.

[29] University College London Calendar 1959–1960.

[30] Ingold, C.K.: Letter to David, F.N. 12 May 1959.

[31] David, F.N.: Letter to University College London vice provost 13 May 1959.

[32] University College London Administrator: Letter to the University College London academic registrar 19 May 1959.

dated it 31 July 1962. She prepared it for the Geophysics Research Directorate, Air Force Cambridge Research Laboratories, Office of Aerospace Research, US Air Force. In the 18-page paper, David proposed a stochastic model for precipitation at two stations that were a specified distance apart. It appears to be related to the work that Neyman and Scott were doing on cloud seeding.

David's international reputation as an academic was growing. As described, she had requested and been given a leave of absence with pay in 1958 for 2 weeks and in 1959 for 5 weeks. The next year, in 1960, she requested another leave of absence with pay, this time for about 6 weeks. She was invited back to Berkeley to give some more lectures, read a paper at Neyman's 5th Berkeley Symposium on Probability and Mathematical Statistics, and do some consulting work for the US Office of Naval Research. David expressed hesitation to her provost about asking for a leave of absence with pay for a third year in a row. She explained to him: "It is, however a great stimulus to me - each time I have been away I have written four papers - and it is a time of year when I can be spared from the department without leaving extra work for other people."[33]

The UCL provost was very supportive of David's request. He wrote: "It is true that you have had leaves of absence in 1958 and 1959, but at the same time the College is very aware of its deep debt to you for all that you have done for the Department and, indeed, for the College more generally. I am sure, therefore, that the College Committee would be most anxious to meet your wishes in this matter."[34] David was approved for a leave of absence from 26 May to 6 July 1960.[35]

David sent a thank you letter to her provost for the leave of absence. She wrote: "It is generous of the college to let me off in this way for three years running and I appreciate it very much." David was so connected to the military given her work for the World War II effort that she would be flown to the United States via the Military Air Transport Service (MATS). She wrote: "I am being flown over by MATS so after my experiences coming back last year - we had three goes at crossing the Atlantic - I hope I get there." David had apparently traded comfort for safety.

In the same letter to her provost, she demonstrated the leadership competency of emotional intelligence by being empathetic and mindful of him. They had both attended a meeting the night before. She wrote: "I would also like to say again how very much members of the non-professional delegation appreciated your meeting and talking with us last evening. Looking back on it I think perhaps I ought to have talked a little more and taken some of it off you. You must have been tired being the focus of concentration for nearly two hours. But I was anxious that they should all talk to you, and I know they felt pleased and satisfied to have done so."[36]

E Pearson was able to make satisfactory arrangements "...for the continuation of teaching and research in the Department during Dr. David's absence" and accordingly approved all three leaves of absence with pay.[37] At the time, the department was offering courses for the B.Sc. (special) degree in statistics, courses for the third year of the B.Sc. (special) degree in math, and for a 1- or 2-year ancillary program in Statistics. The department was also offering short lecture courses for students in other programs.[38]

E Pearson recommended a promotion for Norman Johnson and an increase in salary for Maxine Merrington. Yet, around this time, the personnel action that E Pearson advocated to be taken for David was, in fact, *no action*. This was apparently to help prohibit her from being considered for his position.[39]

---

[33] David, F.N.: Handwritten letter to University College London provost 4 January 1960.

[34] University College London Provost: Letter to David, F.N. 5 January 1960.

[35] University College London Records Department: Approval of leave of absence 2 February 1960.

[36] David, F.N.: Handwritten letter to University College London provost 4 February 1960.

[37] University College Records Department: Approval of leave of absence 2 February 1960.

[38] University College London Calendar 1960–1961.

[39] Pearson, E.: Handwritten notes n.d.

**Fig. 7.2** David at Berkeley with fellow statisticians E.L. Scott, D. Blackwell, and E. Fix (1960). (Image credit: Reprinted with the permission of the Institute of Mathematical Statistics)

David had a problem with her visa this time. She wrote to Neyman:

> My dear J.
> This is just to say thank you very much and to tell you that I think you're terrific. I had a letter from the S. Francisco immigration authorities by the same post as I heard from you. I went to the U.S. Consulate who said they had heard nothing but grudgingly consented to take action. I swore not to blow up re White House and all the rest of it and when I went in yesterday afternoon to pick up my visa all were smiling - they had a cable from Washington and all their red tape was o.k.
>
> L. and K.,
> D[40]

Elizabeth L. Scott ("Betty") and David were friends. Betty was on sabbatical when David wrote to her about plans to return to the United States.

Betty was helping David with the arrangements, and they weren't going well on David's end. David wrote: "I am sorry to cause trouble and can only say I am v. grateful. I will endeavor to say nothing which will contradict anything you have said. I wonder sometimes what it is all about - I really haven't much of a clue - but am content to leave it all in your and Mr. Neyman's good hands. I have made a booking with BOAC [British Overseas Airways Corporation] for June 17th (if that is a Saturday). I suppose I would like to take this up sometime about May 20th (or cancel it) but maybe I will hear something before then." (Fig. 7.2)

David wrote to Betty from the Alps. She had been there for 2 weeks, and she told Betty she didn't want to go back to London. "It hasn't been a v. good year," she told Betty. David was having some trouble with her eye, and the Alpine climate with its clear air and 70° temperature suited her "admiraly." She went on to say "If I could find a suitable place here [in the Alps] I would retire from the University as soon as possible and stay here for at least six months in the year. . . on my walks I look at 'cabins' and maybe I will find one."[41]

---

[40] David, F.N.: Handwritten letter to Neyman, J. 19 May 1960.

[41] David, F.N.: Handwritten letter to Scott, E.L. 10 April 1961.

In a set of notes for the 1960–1961 session, E Pearson wrote that "As soon as my successor in the Chair is appointed, I should naturally discuss the whole portion both with him and the Provost." Regarding David, he wrote "Dr. David's position and future being a little uncertain, I have made no comment on the question of special personal salary."

E Pearson was closer to retirement now and getting ready to shift responsibilities to his successor.[42] In 1960, he recruited Bartlett as chair of the department of statistics. E Pearson had previously recruited him to his department as a faculty member in 1933, while Neyman was there, but Bartlett did not stay. He went as a researcher into the Ministry of Supply during World War II and then as a professor to the University of Manchester. So this was Bartlett's second appointment at UCL. This time he would stay at UCL for 7 years, until 1967.

David had been passed over by her mentor, E Pearson.[43]

Fortunately, David liked and respected Bartlett. When he arrived, the two of them talked. The conversation between them went sufficiently well that David decided to stay at UCL, and she put aside negative feelings [4].

In 1962, 2 years after Bartlett became the new head of the statistics department, David was promoted to the rank of professor at UCL. Bartlett's title was professor of statistics and head of department. David's title was professor of probability and statistics.[44] David was "quite pleased" to be named to the professor rank [4].

David's professor appointment was one that many thought she deserved, but it had been withheld by E Pearson due to his fears that she might succeed him as department head. Consequently, she had to wait for someone else to nominate and support her to receive the professor rank.

UCL had previously appointed only one woman to the rank of professor. This was Dame Kathleen Lonsdale, D.B.E., D.Sc., F.R.S. (b.1903–d.1971). A chemist, she was a crystallographer who discovered the structure of benzene and hexachlorobenzene.

A year after David's appointment to the professor rank, in 1963, Berkeley Statistics Professor Evelyn Fix (b.1904–d.1965) visited London on a sabbatical. The UCL department appointed her, with the title Honorary Research Associate.[45] Fix had earned her PhD degree in statistics at Berkeley in 1948 under Neyman on distributions that lead to linear regressions. She contributed to the World War II effort as a research assistant in Neyman's Berkeley Statistical Laboratory which was doing projects as part of the National Defense Research Committee's Applied Mathematics Panel. An obituary said this:

> The war years were hard. They brought difficult, but occasionally interesting problems. Every demand on the Laboratory emphasized urgency; the practical solutions of the problems meant the determination of optimal plans, the probability of success of this plan, of that plan, numbers and more numbers. High speed computers did not exist; all the numerical work had to be done on disk calculators consuming much time and effort. Endowed with unusual energy and a special spirit of getting the job done and done right. Evelyn [Fix] spent days and nights at her machine, aided by a group of students and some faculty wives, so that the needed results could be transmitted on time, usually to New York but occasionally directly to England. During these years Evelyn also continued her own studies and lectured to students. [6]

Fix was appointed to the Berkeley mathematics faculty in 1950. She was one of the two women among the founding members of the Berkeley statistics department in 1955. She was promoted to the rank of professor of statistics in 1963. She had a doctoral student, William Robert Gaffey, who did his dissertation on a problem of within family contagion [7].

[42] Pearson, E.: Notes session 1960–1961.

[43] Handwritten notes prepared around 1967. Note: E Pearson refers to David as his mentee.

[44] University College London Calendar 1964–1965.

[45] University College London Calendar 1963–1964.

Fix is best known for her paper with Joseph Hodges, Jr. on discriminatory analysis published in 1951.[46] This paper defined the nearest neighbor rule which was a forerunner to the k-nearest neighbor (k-NN) algorithm in machine learning. Fix was elected to be a Fellow of the Institute of Mathematical Statistics. If statisticians as a professional group are known to be generous with their statistical help, Fix stood out even among her colleagues in this regard.

Over the years, Fix and David worked together on statistical problems in biology and health. They were companions and bought a house together at 156 Highland Blvd. in Kensington, CA, which is a suburb of Berkeley. Fix would only live for two more years after her sabbatical at UCL, dying of a heart attack in 1965 at age 61. Fix's Wiki page described David and Fix as life partners.

In May of the year Fix died (1965), David was in Berkeley again. She wrote to Professor of Human Genetics Lionel Penrose, then director of the Galton Laboratory at UCL. Calling him her "favorite genius," she wrote him a special letter urging him to accept Neyman's invitation to the next Berkeley Symposium. She thought Penrose and his staff could come up with an interesting session about "chromosome stuff" that they had been working on over the past few years. David further thought that he would like that fact that "(mostly) people who are trying to work on the applied side" would be there, and the symposium would have an English flavor since Bartlett and Barton were definitely coming.

David told Penrose that she would be back at UCL in time to "open the shop for the new session." She didn't want to return so soon, but she needed to in order to help Bartlett out with his own plans for a leave of absence. She told Penrose "I contemplate the London winter with something less than joy." She also expressed a hope to Penrose that he would give her some office space in his UCL lab and that he would continue to give lectures in the statistics curriculum.[47]

Over 20 years after K Pearson's death, E Pearson continued to be protective of his father's legacy. David wrote to Neyman:

> I hear from Louise that you have been misbehaving again. I hope you are better by now. E.S.P. [E Pearson] is alright I think, but the operation has aged him quite a bit. We had a devastating afternoon last week trying to clear up K.P.'s [K Pearson's] manuscripts. I threw a lot away and later found him [E Pearson] going through the waste paper basket and taking them all out again. I gave up at this point. He will have to do something about them during the next six months. Maybe it is best to wrap them up and put them in the basement. What is really required is a good fire.[48,49,50,51]

In the mid-1960s, both E Pearson and Neyman were relying on David to review and comment, or otherwise provide advice, on their historical writing. E Pearson's daughter Sarah was in Berkeley, and Neyman introduced her to David and Joyce. Continuing to be collegial, E Pearson expressed to Neyman the hope that "David was finally happy about the Tables of Symmetric and Allied Functions; the Press sent her an advance copy of the book before she left, but I did not see her afterwards."[52]

---

[46] Fix, E. and Hodges, J. Jr.: Discriminatory analysis. Nonparametric discrimination: Consistency properties. USAF School of Aviation Medicine, Randolph Field, Texas (1951).

[47] David, F.N.: Handwritten letter to Penrose, L. 20 May 1965.

[48] Pearson, E.: Handwritten letter to Neyman, J. 9 July 1965.

[49] Neyman, J.: Letter to Pearson, E. 12 July 1965.

[50] Pearson, E.: Letter to Neyman, J. 5 August 1965.

[51] Neyman, J.: Letter to Pearson, E. 16 December 1966.

[52] Pearson, E.: Handwritten letter to Neyman, J. 2 April 1966.

David informed Neyman she would be arriving to Berkeley on December 15th by Pan-American, via Seattle and Portland.[53] She told Neyman to "keep his hair on," which is a British way of saying that he should "stay calm."[54]

To summarize, records indicate that David was in Berkeley as a visiting professor in 1958, 1959, 1960, 1965, and 1966. She had one foot solidly out of the UCL door.

## 7.3 Three Books on Probability and a Festschrift

David's last decade at UCL (1957–1967) was a productive span of years for her. She published four books.

The first of the four was the book, *Games, Gods and Gambling: A History of Probability and Statistical Ideas*. This book was initially published by Charles Griffin and Co. Ltd., London, and Hafner Publishing Company, New York, in 1962. It was published again by Dover Publications in 1998. We have already described this book in some detail.

The second book that David published during this period was with Barton (1962), titled *Combinatorial Chance* [8]. The third book was with Kendall and Barton (1966), titled *Symmetric Function and Allied Tables* [9]. The books were published by Hafner and Cambridge University Press, respectively.

David and Kendall wrote the 1962 book on combinatorial analysis as a way of preparing for the 1966 book on symmetric functions. The two books therefore could have been regarded as companions, although they were designed to be used independently. The first book considered combinatorial methods for their own sake, and the second coupled them with statistical methods.

First we will examine the combinatorial chance (1962) book. David and Barton read widely in the literature to develop this book. They wrote it at an advanced level with the aim to provide a compendium of combinatorial methods that would be useful to statisticians. Statisticians at the time were using combinatorial chance more frequently in mathematical statistics. Reviewers agreed that there was a lot in this book that would be of great use to statisticians. Topics included generating functions, card problems, dice problems, derangement and matching, randomization, occupancy problems, run problems, permutations, and more.

David had worked on combinatorics for some time, but then she got bored with it. She took Barton on as a student, and then he worked with her to finish the book. She said "[Barton] did all the fancy work, proceeding to limits and things like that. He was a good chap." [4]

Coming on the heels of the *Games, Gods and Gambling* book, David and Barton infused history and their own recent work into the *Combinatorial Chance* book. The authors even dedicated the book to Abraham de Moivre (b.1667–d.1754), based on his works *De Mensura Sortis* (1711) and *Doctrine of Chances* (1718, 1738, and 1756.) There were a lot of equations, tables, and numerical examples in the book.

David was a learned person and called up wonderful quotations. At the front of this book, she used "Heureux qui, comme Ulysse, a fait un beau voyage" which translates to "Happy is he who, like Ulysses, has made a beautiful voyage." Perhaps David felt that she and Barton had made a marvelous journey of their own by writing this book. David followed all of this by, in capital letters, "To His Illustrious Shade," which we believe is a dedication to the sailor Ulysseus's shade (i.e., ghost) in the Greek underworld.

---

53 David, F.N.: Handwritten letter to Neyman, J. 29 November 1966.

54 David, F.N.: Telegram to Neyman, J. n.d.

Reviewer W. G. Gilchrist (Sheffield City Polytechnic) opined that the delineation of combinatorial methods in the book was both "comprehensive" and "clear." He singled out the first chapter for praise: "This chapter is devoted to a number of simple examples illustrating the topics and methods to be discussed in the rest of the book. One wishes that more authors would lead the reader into their subjects by first giving him some basic comprehension and insight in this way." But then he criticized the authors for leaving the applications of the methods to the subsequent book on symmetric functions. He ended his review with these positive words: "Nevertheless, leaving this debatable criticism aside, this is a book to be recommended not only to those particularly interested in its topic, but to anyone who enjoys a clear presentation of some very elegant methods." [10]

John Riordan (Bell Telephone Laboratories) was critical and pointed out "numerous sins against unity, coherence and emphasis" as well as some "allergies." He thought some of these problems might have resulted from the "well-known difficulties of co-authorship." He concluded his review by saying "And despite all criticisms, it may be that the working statistician on some occasion will find in this book exactly what he needs." [11]

There was a shorter notice about this book in the journal of The London School of Economics and Political Science. It called the book "comprehensive and straightforward." The unnamed reviewer saw the book, due to the necessarily heavy use of algebra, as more of a reference book for statisticians than a text book, and that it would be most useful to third or higher year students of mathematical probability theory [12].

Reviewer A. Huitson (Sheffield Polytechnic) wrote about how few new books on probability are published and why. This reviewer thought that "the authors are to be congratulated on having produced a notable work." They went on to say that "... most applied statisticians will benefit either directly or indirectly from this work by getting to grips with one of the fundamental properties of statistical data, namely, its combinatorial feature... The competent statistician will find the work interesting and useful in his practical problems... the book will be a credit to many bookshelves in more ways than one." [13]

D.G. Beech (British Ceramic Research Association) would have preferred the first book that included the coupling with statistical methods: "... it is a pity that the attempt was not made...." Instead, the authors produced a book that demanded "a fair degree of esoteric knowledge." The reviewer concluded with this statement: "Despite all this, the authors are to be congratulated on a very substantial piece of work on a fascinating subject. Those whose interests lie in this field would be well advised to tackle this book provided they have first mastered a preliminary textbook." [14]

Reviewer Edward B. Roessler (University of California - Davis) added that the book was clearly written and carefully illustrated by well-chosen examples [15]. John E. Walsh (System Development Corporation) pointed out that the material should be especially useful in the area of nonparametrics. He thought the reference lists were "somewhat skimpy" but that "on the whole," it "seems to be a worthy contribution to the literature of probability and statistics." [16]

E. Sverdrup (University of Oslo) articulated the purpose of the book: It was "to present a unified mathematical treatment of the variety of intricate problems in combinatorial theory, problems which are found under different disguise in various fields of application." He recognized a real need for combinatorial methods: "as, for instance, psychological research, where only ranking order and not absolute measurements are possible; time series analysis with investigations of runs up and down, etc.; or, industrial quality control where comparisons, not measurements, are deemed convenient." He opined that "The great advantage of the present treatment of the subject as compared with its treatment in statistical literature is, that whereas the latter deals with the problems as they arise in their statistical context, the present treatment is systematic and treats related problems under one heading." He went on to say that the presentation was very clear and that "it will be difficult to find any statistician or probabilist to whom the present book will not be useful and who will not find many results in it which were unknown to him." [17]

E. N. Gilbert (Bell Telephone Laboratories) pointed out that the David and Barton book covers elementary parts of probability theory—the part that has to do with counting cases in probabilistic terms—and that "elementary" should not be confused with "simple." He especially liked Chap. 5: It "introduces two of the book's best features. One is the generating function identity relating moments (which are often easy to find) to probabilities. The authors achieve some neat derivations later on by working from moments to probabilities. Secondly, Chap. 5 discusses limit theorems in preparation for numerous Gaussian and Poisson approximations which they give later on." The main mathematics that the reader needs to know to approach this book, according to Gilbert, is infinite series. He also thought the reader should know some elementary combinatorial analysis. He concluded: "The book's somewhat disjointed organization and its lack of exercises for the reader may limit its use as a textbook. Its usefulness as a reference book is enhanced by its large number of problems of statistical interests, by its many tables, and by a five page glossary of notations and abbreviations most of which are followed consistently throughout the book." [18]

W. Feller (Princeton University) noted that "few people have an understanding of the peculiar fascination of combinatorial problems for those who live with them. The authors belong to the latter class; they feel that combinatorics can be fun, and they try to communicate this feeling." He pointed out that the book includes a broad range of methods, from simple arrangement problems to statistical tests that are intricate. Feller thought that the plan for the book was "excellent, and affords an opportunity to introduce some of the statistician's standard tools almost playfully by informal discussion and intuitive examples." He thought, however, that the plan wasn't well executed: "This may not seriously affect the potential user or experienced statistician who will find a wealth of detailed information and worked-out formulae. However, an outsider may have a hard time to orientate himself, and a beginner will experience difficulties even in the elementary chapter written for him...." Feller concluded with the note that "It is to be hoped that a revision may render the book truly enjoyable." [19]

Barton declared that, when it came to combinatorial problems of probability, David "...had both technical fluency and insight." Over a 17-year period starting in 1951, she either wrote or cowrote about 30 papers on different combinatorial problems and applications [20].

Now we will examine the symmetric functions (1966) book. E Pearson asked David, Kendall, and Barton to collaborate on a volume of combinatorial tables, which became this book. All three collaborators were already well known for their published work on combinatorics, and the book turned out to be more expansive than the original suggestion [21]. David said that she was interested in combinatorics because she was "lazy" and wanted to find quicker ways of doing things [4].

The symmetric function book is primarily a set of 49 major tables, some of them published for the first time. In the preface to the book, David explained that she and Kendall wanted to make combinatorial statistical procedures easy to use, and they accordingly created statistical tables. Reviewer F. Gobel (Mathematical Centre in Amsterdam) commented that "The authors have devised a number of novel ways of organizing and presenting the tables." [22] Reviewer D.S. Tracy (University Windsor, Ontario) called the format of the tables "remarkably good" and said that "the authors have organized them rather ingeniously." [23]

Following the format of David's first book, on correlation, the tables were preceded by a detailed and self-contained introduction to the theory of symmetric functions and their statistical applications. The introduction brought together contributions that were, until then, scattered across the literature. Together, the introduction and tables made for a "definitive compilation," although Gobel's review noted a number of inaccuracies in the book [24].

It is no surprise that David's preface started with a brief history. It was quite complimentary of R.A. Fisher. However, she pointed out that the idea of augmented symmetric functions was her and her co-authors' original contribution. They summarized all of the techniques they themselves found useful, and by doing so, they paved the way for future contributions by others. The book was described as "elaborate" and "attractively printed."

Reviewer N.A. Rahman (University of Leicester) remarked that "One of the continual fascinations and difficulties of mathematical statistics is that, quite frequently, the newer developments of its theory bring into use a relatively unfamiliar branch of pure mathematics... the essential point is that, in a discipline so rapidly developing as mathematical statistics, it is incumbent on the statisticians themselves to provide the mathematics to their future colleagues which is most germane to their science." He then lauds the authors for bringing together what he calls a "neglected but pervasively used" area of science.

Rahman called the book "sturdy scholarship," the introduction "excellent... alone an admirable monograph...," and the tables "useful... produced with meticulous care...." He said the authors used "judicious care" in appealing to people both with and without a background in combinatorics, and they put in "extensive work" to produce this and the earlier volume. The reviewer recommended that a beginner should start with the earlier volume: "... [he] will be well rewarded for his pains; and pains there will be, since the subject demands much specialized ability and manipulative skill which is far too often neglected in mathematical education." The reviewer further hoped that the two books would stimulate the general study of combinatorial analysis in statistical education.

It is noteworthy that this reviewer, Rahman, gave credit to Kendall and E Pearson, even though David was the first (senior) author and wrote the book's preface. We see this oversight as an example of how women in science can be diminished [25].

Then reviewer B.E. Cooper (ICL Dataskil, Ltd., in Reading, UK) made an ominous comment:

> In preparing this review, I cannot dismiss the thought that we must be approaching the end of an era in Statistics. Tables have been produced to reduce the burden of applying statistical procedures and they have taken a vital place in the development of new methods. However, now that computers are widely available, and as remote access becomes more nearly the normal facility, the need will change from published tables to algorithms for computing single entries. In this respect the comments made in introductions to tables often have little relevance. Most tables have been built up from one corner of the table using recursive relations between rows and columns and such computing methods are not usually the best method of computing single entries. For example, in the introduction to Table 4.2 three methods of computing are discussed. Of the three only one is non-recursive and therefore available for computing a single entry. However, the non-recursive method is not described being dismissed with the comment that "it is not less-laborious than the previous method." Thus it is expected that tables of distributions will be replaced by algorithms for computing the appropriate tail probability, and that relations such as those given in Part I of this volume will be obtained by computer interrogation from the remote console. In looking to the future, however, I do not mean to detract from the hard work of the present authors nor from the usefulness of this volume. [26]

First with her book on correlation, and later with her book on symmetric functions, David had worked hard to create tables that would make statistics easy to apply. But now the world of statistics was changing. David's skill and expertise for creating tables were being replaced by high-speed computers.

David's next book was not tables, but a Festschrift for her close colleague and friend Neyman [27].[55] She had the distinction of being Neyman's first postgraduate student at UCL. David wrote the following in the Foreword to the book:

> It had long been the wish of us here at University College, London, to express in some way the esteem in which we hold Jerzy Neyman as a statistician, the pleasure which we have in him as a friend and the pride which we

[55] Note: A Festschrift is a volume of papers, written by different scholars (colleagues, former students, friends), in order to honor a particularly eminent scholar.

> have in recalling that it was while he was here that he published the theory of confidence intervals and, with E. S. Pearson, the theory of testing a statistical hypothesis. Opinions differed as to how these various appreciations should be expressed and it was not until the arrival of Evelyn Fix on a sabbatical year at the College that it became apparent that a book of essays and research papers might be the appropriate token offering.

Evelyn Fix not only had degrees in mathematics and statistics, but she also had a degree in library science and was instrumental in organizing the proceedings of the Berkeley Symposia on Mathematical Statistics and Probability [5] (p. 176). Fix and David shared a deep knowledge of books and were naturally inclined at thinking together about a book to honor Neyman. David was the editor of the volume. Fix was sufficiently helpful that David acknowledged her on the cover page: *Assisted by Evelyn Fix*.

The book was meant to honor Neyman on his 70th birthday. Eminent scholars from around the world wrote research or review papers for the book. Chapter authors included E Pearson, M.S. Bartlett, J. Berkson, D.R. Cox, H. Cramer, D. Dugue, R. Fortet, U. Grenander, J.M. Hammersley, Koichi Ito, D.G. Kendall, L. LeCam, P. Levy, P.A.P. Moran, Thornton Page, George Polya, C. Radhakrishna Rao, Alfred Renyi, Sixto Rios Et Ildefonso Yanez, L. Schmetterer, Herbert Solomon, Charles Stein, P.V. Sukhatme, Herman Wold, D.E. Barton with David, and David with Evelyn Fix.

There were 28 chapters addressing a wide range of topics across the fields of statistics and probability. These included many valuable references. Reviewer and Statistics Professor T.A. Bancroft (Iowa State University) highly recommended the book as an excellent reference book for statistics students and faculty alike [28].

As she wrote in the Foreword, David presented the book to Neyman "with affection and appreciation of a lifetime spent, and still continuing to be spent, in the service of statistics." Reviewer J. Gani (University of Sheffield, UK) wrote about the "affectionate esteem" in which people across the globe held for Neyman. He discussed some of the chapters and concluded that "Many of the authors have indicated in detail the relation of their work to Jerzy Neyman's research. I can think of no more fitting tribute to his assured place in the annals of statistical history." [29]

The first chapter of this book was special, because E Pearson told the story of his important collaborations with Neyman which determined how we test statistical hypotheses. He concludes with a personal reflection that is truly inspiring:

> I can recapture some of the intellectual thrill of that time; the exhilaration which goes with the belief that one is chipping away along the fringe of the unknown. What value is or will be placed on our joint work suddenly seems relatively unimportant. It is the experience within oneself and the joint friendship which have really mattered in life. [27]

E Pearson's remarks can be viewed as a prayer for the scientist. They indicate that in spite of his lack of support in promoting David or recommending her as his successor, he had some redeeming qualities as a mentor to her. However, David had cultivated reciprocal professional partnerships around the world that opened doors to greater opportunities.

David was numbering her days at UCL.

## References

1. Subramanian, S.: A Dominant Character: The Radical Science and Restless Politics of J.B.S. Haldane. W.W. Norton and Company (2020)
2. The Cambridge Dictionary: Temperament (2022) https://dictionary.cambridge.org/us/dictionary/english/temperament Accessed 7 March 2022
3. Lehmann, E.L.: Reminiscences of a Statistician: The Company I Kept. Springer, pp. 116–119 (2008)
4. Laird, N.M.: A conversation with F.N. David. Statistical Science 4(3), 235–246 (1989)
5. Reid, C.: Neyman-From Life Springer-Verlag, New York (1982)

6. Neyman, J., Hodges, J.L., Jr., LeCam, L., Lehmann, E.L., and Scott, E.L.: Evelyn Fix, Statistics: Berkeley. (1965) https://statistics.berkeley.edu/about/biographies/evelyn-fix Accessed 3 August 2023
7. University of California—Berkeley: People in the Department of Statistics (2022) https://statistics.berkeley.edu/people/william-robert-gaffey Accessed 6 April 2022
8. David, F.N.and Barton, D.E.: Combinatorial Chance, Hafner Publishing Company, New York (1962)
9. David, F.N., Kendall, M.G., and Barton, D.E.: Symmetric Function and Allied Tables, Cambridge University Press (1966)
10. Gilchrist, W.G.: Review of Combinatorial Chance by David, F.N. and Barton, D.E. Journal of the Royal Statistical Society, Series A (General) 125(4), 639–640 (1962)
11. Riordan, J.: Review of Combinatorial Chance by David, F.N. and Barton, D.E. The Annals of Mathematical Statistics 33(4), 1477–1479, December (1962)
12. The London School of Economics and Political Science: Review of Combinatorial Chance by David, F.N. and Barton, D.E. Economica, New Series 29(115), 332, August (1962)
13. Huitson, A.: Review of Combinatorial Chance by David, F.N. and Barton, D.E. Journal of the Royal Statistical Society, Series D (The Statistician) 12(4), 333 (1962)
14. Beech, D.G.: Review of Combinatorial Chance by David, F.N. and Barton, D.E. Journal of the Royal Statistical Society, Series C (Applied Statistics) 11(3), 211, November (1962)
15. Roessler, E.B.: Review of Combinatorial Chance by David, F.N. and Barton, D.E. The American Mathematical Monthly 70(3), 351, March (1963)
16. Walsh, J.E.: Review of Combinatorial Chance by David, F.N. and Barton, D.E. Journal of the American Statistical Association 58(301), 251–253, March (1963)
17. Sverdrup, E.: Review of Combinatorial Chance by David, F.N. and Barton, D.E. Review of the International Statistical Institute 31(1), 143–144 (1963)
18. Gilbert, E.N.: Review of Combinatorial Chance by David, F.N. and Barton, D.E. Technometrics 5(2), 289–290, May (1963)
19. Feller, W.: Review of Combinatorial Chance by David, F.N. and Barton, D.E. Biometrika 50(1/2), 225, June (1963)
20. Barton, D.E.: Obituary: Florence Nightingale David Journal of the Royal Statistical Society A 157(2), 299–303 (1994)
21. Cooper, B.E.: Review of Symmetric Function and Allied Tables by David, F.N., Kendall, M.G., and Barton, D.E. The Mathematical Gazette 56(395), 69–70, February (1972)
22. Gobel, F.: Review of Symmetric Function and Allied Tables by David, F.N., Kendall, M.G., and Barton, D.E. Biometrika 54(1/2), 335–336, June (1967)
23. Tracy, D.S.: Review of Symmetric Function and Allied Tables by David, F.N., Kendall, M.G., and Barton, D.E. Review of the International Statistical Institute 36(3), 351–352 (1968)
24. J.W.W.: Review of Symmetric Function and Allied Tables by David, F.N., Kendall, M.G., and Barton, D.E. Mathematics of Computation 21(98), 269–270, April (1967)
25. Rahman, N.A.: Review of Symmetric Function and Allied Tables by David, F.N., Kendall, M.G., and Barton, D.E. Journal of the Royal Statistical Society, Series A (General) 130(2), 256–257 (1967)
26. Cooper, B.E.: Review of Symmetric Function and Allied Tables by David, F.N., Kendall, M.G., and Barton, D.E. The Mathematical Gazette 56(395), 69–70, February (1972)
27. David, F.N.: Editor, Research Papers in Statistics: Festschrift for J. Neyman, John Wiley and Sons (1966)
28. Bancroft, T.A.: Review of Research Papers in Statistics Edited by David, F.N. Technometrics 11(4), 838–839, November (1969)
29. Gani, J.: Review of Research Papers in Statistics Edited by David, F.N. Review of the International Statistical Institute 35(3), 322–323 (1967)

# Part V

# In the New World

# Arriving (1967–1974) 8

*I am grateful to the College and to the University for putting up with me for thirty-two years... But I am 58 in August and the grass is greener on the other side of the hill.*

David, F.N.: letter of resignation from UCL (1967)[1]

## 8.1 Separating from UCL

Heading into 1967, David remained in the United Kingdom. It had only been about a year since Evelyn Fix's death, and one of David's associates wished for "all that melancholy business is now behind you."[2] Already David must have known that her statistics department head at UCL, Maurice Bartlett, was making plans to move to a biomathematics faculty position at the University of Oxford. This once again made her future at UCL precarious, and she was preparing to make a move.

David recognized that opportunities in the 1950s and 1960s for a statistics professor in the United Kingdom were very limited. She had already seen a number of UCL statistics faculty members emigrate to the United States, including Herman Otto Hartley in 1953 and Norman Lloyd Johnson in 1962. The United States was the other side of the hill where the grass was greener.

By February 1967, David knew she was going to move from UCL across the Atlantic to the University of California—Riverside (UCR). Statistics at UCR was still being taught in the mathematics department. The plan was to have a new department of statistics with a menu of statistics degrees at all levels: an undergraduate major, a Master of Science degree, and a PhD degree.

Morris J. Garber (b.1912–d.2004) was already on the faculty at UCR. His PhD degree was in genetics, so he was perfect for a new department of statistics which would focus on biological and agricultural applications of statistics. Garber's official University of California obituary reported that he was instrumental in the founding of the statistics department at UCR: "In addition to bringing UCR into the age of computing, Garber was instrumental in administering the Biometrical Laboratory, which gave rise to the Biometric Project, and in developing the proposal for the Department of Statistics and the proposal for a building to house the Department of Statistics and the Computing

[1] David, F.N.: Letter to University College London principal 8 May 1967.

[2] John (in Gainesville, FL): Letter to David, F.N. 2 February 1967.



A. L. Golbeck, *Florence Nightingale David*,
https://doi.org/10.1007/978-3-031-77768-4_8

Center (the Statistics-Computer Building)." [1] David was recruited to grow and shape the department that Garber proposed.

David's closest colleagues around the world knew she was leaving UCL and going to UCR. They wished her "much luck and happiness." Somehow her plans had gotten out and were being whispered about in the United States, even though David had wished otherwise. Fortunately she had already informed her UCL department head Bartlett at the beginning of October, and she informed her provost Noel Annan in December that she was considering an offer from UCR.[3]

The move to California would be in the midst of a budget downturn in the state. Ronald Reagan was the new governor. In one of his first acts, he cut the University of California's budget by 30 percent. This would undoubtedly impact David's plans to grow a statistics department at UCR. The question was how much.

A close colleague in Florida shared this perspective: "The clashes between education and politics in [Florida] are regrettable but it does seem now that they will be worse in California." The same person closed his chatty and affectionate letter by sending his regards to David's friend Joyce Townend,[4] who David had known for over 30 years at UCL, saying "I hope she's going to California with you—good friends are a treasure."[5]

Another colleague was more reassuring, saying he hoped David would not "let the educational news from California stop or deter you in your new position at Riverside. There is an air of 'going backwards' in the whole setting, but on the realistic side I don't see how the trend can be anything but expansion as the State and its needs grow."[6]

Some close colleagues saw possible opportunities for themselves or their students with David at UCR. Prem S. Puri (b.1936–d.1989), for example, hoped to meet up with her in the summer of 1967. Neyman had recruited Puri to a summer appointment at Berkeley, and Puri wanted to discuss opportunities with David at UCR.[7]

By the end of April 1967, David had received a formal offer from UCR. She wrote to her provost at UCL:

> We have had conversations about what I should do as the result of a formal offer made to me by the University of California. I write to ask that the college allow me a year's unpaid leave of absence in order that I may spend some time at the University of California Riverside campus. This is one of the new campuses of the University of California. They are anxious to develope [sic] a department of statistics of the English type. I discussed what was required with the biological and agricultural sciences dean when I was there last week and I think I can do a good job both for Statistics generally—which right now is rather in the doldrums—and for the biologists. I would like the leave of absence to run from October 1st, 1967 to September 30th, 1968.[8]

A leave of absence would have been the professionally safest way for David to make her move, because if things failed to work out at UCR, then she would have had a position to go back to at UCL. It would have also been helpful to the statistics department at UCL, because it would have allowed for a valuable asset (i.e., David) to possibly return to their ranks.

But there was a new sheriff in town. The new provost was Noel Annan (b.1916–d.2000). Being a former British military intelligence officer and a baron, he brought the rigidity of military command and the entitlement of royalty rank to his position. Annan answered David with a crusty message:

---

[3] David, F.N.: Letter to University College London principal 8 May 1967.

[4] Note: Townend and David apparently also had some professional dimensions to their relationship. For example, David credits Townend for drawing the figure on the paper that David published on Neyman's smooth test.

[5] John (in Gainesville, FL): Letter to David, F.N. 2 February 1967.

[6] Solomon, H.: Letter to David, F.N. 13 February 1967.

[7] Puri, P.S.: Letter to David, F.N. 18 April 1967.

[8] David, F.N.: Letter to University College London provost 28 April 1967.

> I am somewhat embarrassed by your request in that I'm told that, disregarding brief absences, you have been granted the following leaves of absence during the past ten years:
>
> 1958–9 half a term, with pay
> 1959–60 about half a term, with pay
> 1961–2 whole session, without pay
> 1964–5 whole session, without pay
> 1965–6 one term, with pay
>
> In normal circumstances, I don't think it would be possible to give a further leave of absence for next year.
> It is, indeed, particularly difficult to do so in that, as you probably know, Dr. Lewis has been granted leave of absence for the whole of next year—and we've not yet appointed a new Professor of Statistics.
> If we were unable to make an appointment to take effect from 1 October 1967, I would regard it as imperative that you stayed at University College to act in Professor Bartlett's place. If we were able to make an appointment, it would, of course, be a matter of decision for you to see if you wanted to stay and work with the new man who would, I am sure, benefit from your help. I would fully understand and respect your feelings if you found yourself unable to work with the new man, but then I think a different situation would have arisen which we would have to discuss.[9]

Given the gender reference ("...work with the new man..."), the provost telegraphed that the appointment would not go to David and also that it would go to a man. However, as he explained to David, with disrespect, that she would be needed to fill in until a man was found, it was as if David was an inherent problem rather than an asset.

A normal reaction to a letter like the provost wrote would prompt indignation on the part of the recipient, and it did. David quickly resigned from her position at UCL, effective 30 September 1967. This was a few days after the university deadline to resign, but it turned out that when she last spoke to the provost (which was in December 1966[10]), he told her to request a leave of absence. She wrote: "I have made my plans for next year on the assumption that [the leave of absence, which you suggested] would be granted and it is too late in the year to change them now."

David added: "I agree I have had a good deal of latitude regarding leave of absence. Maybe its a good thing now to make up my mind."[11] Then she made up her mind, and she resigned.

The provost's reply to David's resignation was arrogant, yet supportive:

> I am sure you are right to have taken the decision to resign. In all our conversations I have felt that even if this vacancy in the Chair had not come up, you had gradually got yourself in a frame of mind when you thought you ought to move. As you say, the time has probably come when you have got to take the plunge.
> It is singularly inappropriate that having been here only six months I should be the one to write and thank you for all that you have done for the College. But even though I have been here for so short a time, I have learnt with what affection and respect you are held by all those who have given long service to the College. Thank you very much for everything you have done. I am sure California will be a success—you have been so sensible in testing the climate there before making up your mind.[12]

Yes, being passed over can cause a talented and accomplished academic to move to a new university.

When David resigned, she asked to keep her apartment at 28 Torrington Square. She would continue to pay the rent. This was, she said, because she had an "aged mama" who might need her assistance, and so she might need to return to "see about her." It must have been difficult for David to decide to move to another continent, given that she would have to leave an aging mother behind. David also asked to be able to keep some of her furniture. She ended up paying UCL 117 GBP for two desks to take with her to America.[13]

---

[9] University College London Provost: Letter to David, F.N. 4 May 1967.

[10] University College London Principal: Letter to David, F.N. 8 May 1967.

[11] David, F.N.: Letter to University College London provost 8 May 1967.

[12] University College London Provost: Letter to David 8 May 1967.

[13] University College London Finance Secretary: Record 21 June 1967.

David wrote: "I am grateful to the College and to the University for putting up with me for thirty-two years (39 if I count my student days). But I am 58 in August and the grass is greener on the other side of the hill."[14]

UCL accepted David's resignation. They congratulated her on her appointment to the University of California, and they thanked her for her many years of service to UCL.[15] She had been a good citizen of her department, having helped develop it "...as a centre both of research and of post-graduate studies while maintaining an excellent undergraduate degree programme." And she had been a good citizen of her university, having served on the academic council and the university senate [2]. The UCL principal even told David he envied her move to California.[16] It is fitting that the final approval was recorded on the 4th of July, the day that the United States celebrates its independence from Great Britain.[17]

The time of David's departure came quickly upon the UCL statistics department, as well as upon David. The secretary thanked David for all she had done for the college and, both personally and on behalf of the college, wished her "every success and happiness".[18] David worked with her accountant to get her income tax documents in order and her life and other insurance policies and pensions. There was a lot of paperwork, and it took many months. The policies considered her departure from UCL to be a retirement.[19]

David had an affiliation with UCB in the summer before heading to Riverside where she used UCB statistics department stationary. In the summer, she arranged to continue to receive her correspondence at the 28 Torrington Square London address.[20]

David had already secured a house in California, which she had previously purchased and shared with Evelyn Fix. It was at 156 Highland Boulevard, Kensington, Berkeley. David thought about the weekly commute that she would be taking between Berkeley and Riverside. She wrote: "Curiously enough although the distance is about ten times what it is between London and my place in Sussex the door to door time is just about the same."[21]

In 1967, the Royal Statistical Society asked its council members and honorary fellows to submit biographical information about themselves. They planned to put the information into an archive for future reference. E Pearson wrote a note, which was marked as having never been sent:

> I am one of the few persons still alive who witnessed at first hand the shift across from the first period of modern mathematical statistics strongly influenced by my father, to the second period in which R.A. Fisher was dominant.
>
> Not being a really expert mathematician, my interests tended to be attracted by applications and to some extent new applications which I was in part responsible for, e.g., in industry and in war time work within the Services.[22]

These paragraphs could just as well have been written by (or for) David, with two exceptions. One is that many would assert she was an expert mathematician. Another is that she did new applications, but they were in government work (agriculture and specifically forestry), rather than industry work.

[14] David, F.N.: Letter to University College London principal 8 May 1967.

[15] University College London Senate: Letter to David, F.N. 19 May 1967.

[16] University College London Principal: Letter to David, F.N. 9 May 1967.

[17] University College London Records: Resignation of appointed teachers 4 July 1967.

[18] University College London Statistics Department Secretary: Letter to David, F.N. 7 August 1967.

[19] The Equitable Life Assurance Society: Form.

[20] David, F.N.: Letter to Davidson 1 June 1967. Note: Mr. Davidson was an accountant in Berkeley.

[21] David, F.N.: Letter to Davidson 21 September 1967.

[22] Pearson, E.S.: Handwritten letter to [David, F.N.] 27 October 1967. Note: The letter was marked "never sent."

Also, UCL deemed E Pearson to be appropriate to be the department head of statistics, whereas E Pearson did not deem David to be so.

In 1967, the position for which David had prepared her whole life—the "premier statistics chair in Britain"—went to David Lindley (b.1923–d.2013) when UCL appointed him as professor of statistics and head of department. Either David had been offered the head position and turned it down (this was her version of the events) [3], or else this was the second time David had been passed over, the first being when Bartlett was appointed head. To make matters worse, Lindley was quite junior in age, being 44 years old compared to David's 57. More importantly, Lindley was a staunch Bayesian, whereas David was a resolute frequentist. *The Guardian* explained it this way:

> The contrast between the Frequentist and Bayesian perspectives is deep. The frequentist approach evaluates the accuracy of an estimate of an unknown value in terms of how different that estimate could have been. The Bayesian approach updates personal beliefs about the unknown true value. The former is an objective statement about the values of statistics generated from a given model; the latter is a subjective statement about beliefs. It was this latter perspective that led to De Finetti's famous remark "probability does not exist."
> By the time of the Fourth Berkeley Symposium on Statistics in California (1960), tensions between the two schools were emerging, and some outspoken exchanges took place. When Lindley moved to University College London, the radically different statistical philosophy he brought with him led Pat Rivett to remark that "it was as though a Jehovah's Witness had been elected Pope." [4]

For David, Lindley wasn't just another new department head. He was also an intellectual arch enemy.

We discovered a YouTube video of Lindley that was filmed in 2013, the year he died [5]. Lindley told the interviewer:

> When I was offered the job [as head of statistics at UCL], I said, Is Florence David coming back from her sabbatical? Because, I said to the Professor, if she is, I will not accept the job, because I couldn't possibly work with her, she's a Frequentist to the core! And he looked at me - very puzzled at this astonishing thing - and within a few days - she was in Berkeley at the time - he got hold of her and she said she wasn't coming back. And I accepted the job. And then two staff left, which was great... because I had two slots I could fill with Bayesians.

Lindley's version of the events is that he persuaded "the indomitable" David to make her move to the New World permanent. But we found indication in the archival record that Lindley took too much credit for David's move. We also found indication that Lindley's persuasion might not have been entirely personal, as Lindley reportedly also encouraged the retirement of two other faculty members at UCL "for not being sufficiently Bayesian"!!! [6]

Lindley spent 10 years (up to age 54) at UCL, rather than settling in for the long term. His own account is that he realized the university rules would allow him to retire at age 54, and he found the idea to be attractive; so he did. Another source tells us that the administration encouraged him to retire after there was a disagreement about keeping a skylight open in one of the offices [6]. In any case, he spent the rest of his years doing what a good emeritus professor would do, namely, staying intellectually engaged.

David arrived in the United States to a position as professor of biostatistics, under the agreement that she would grow a department of statistics. She would become one of the first women in the United States to become a chair of a science department at a research university. Three years into her tenure at UCR, she wrote: "There have been no thesis [sic] written in this department with any Bayesian connotations."[23] She was a firm intellectual leader of her department.

[23] Houle, A.: Letter to David, F.N. 12 June 1972. Note: A. Houle was at Universite Laval in Quebec.

## 8.2 Settling in at UCR

David engaged significantly with the US Forest Service, a consultancy that began in 1958 involving the spread of forest fires [3], and that relationship continued even after she eventually left UCR.[24] The Forest Service must have been a comfortable collaboration for her, as she had been located at a forest station in the United Kingdom during the war. Barton pointed out that her later forestry studies "had military implications." [7]

Most of David's forest service work in the United States was focalized at the branch in Berkeley near the campus [8]. Her primary contact there was William G. O'Regan, a doctorally trained expert in forest sampling. He was the director of biometrical studies and services.[25]

In addition to doing her own statistical work, David placed student assistants in employment at the forest service. Many were paid during the summer and at other times. David also made arrangements for some students to pursue permanent careers in the forest service by communicating about visas and getting to Albany, and expressing concern about their futures.

O'Regan told David he thought of London as a cultural center. This amused David, and so she shared a perspective on differences between the English and the Americans, saying this: "Among the liberties cherished by the English people is their liberty of a freedom from culture, (and how avidly they exercise it). Maybe if we write people as peoples this is true in the U.S. too, but maybe not."

David closed a 1967 letter to O'Regan by telling him that UCB couldn't pay for her to teach that summer. She wrote: "From a personal point of view I am relieved since I said I would teach when I wasn't sure about what money I would have coming in, but now I have some (or so I think) I am well rid of the labor. But [the budget] doesn't look so hot from the general University point of view."[26]

Book publishing opportunities continued to be on David's radar, although she did not pursue them. She received an inquiry from the Wiley and Sons Limited publishing company about a draft book that she had written and was sitting on her shelf. Wiley was interested in whether it would fit their interests in the teaching of statistical inference. They wanted lesser emphasis on descriptive statistics than in the usual textbooks and a book that would compete with other statistics books for non-calculus audiences. The fact that David was intimate with the Berkeley flavor of statistics and could express their point of view made her an especially attractive potential author to them.[27]

Another colleague wrote to David about publishing a book,[28] this time with the McGraw-Hill publishing company. The colleague offered to advocate to the publisher on her behalf. The same colleague noted that the Neyman volume was selling well.[29] But 5 years later, sales would be so slow that the publisher would put the Neyman volume "out of print." This had to be very disappointing for David, given the high esteem she held for Neyman and given the powerhouse of contributors to her volume. She bluntly declared to the publisher she was sorry they were "dumping" the book.[30]

Research articles also continued to be on David's radar. She was furthering her work with Barton on the theory of occupancy, which involved models of balls tossed into boxes. In the authors' words,

---

[24] Author not named: Florence Nightingale David 23 August 1909, Journal not named, n.d.

[25] United States Department of Agriculture: Forest Service Organizational Directory January 1963.

[26] O'Regan, W.G.: Letter to David, F.N. 20 January 1967.

[27] Goulding, O.: Letter to David, F.N. 23 January 1967. Note: Ossian Goulding was John Wiley and Sons Limited chief European editor.

[28] Note: We know that this book was a joint venture with the late Evelyn Fix, but it is not clear whether this is the same book as the one that was being solicited by Wiley.

[29] Solomon, H.: Letter to David, F.N. 13 February 1967. Note: Herbert Solomon was at George Washington University.

[30] Redwood, M.: Letters to David, F.N. 23 August 1972. Note: M. Redwood was John Wiley and Sons Limited managing editor.

these papers "considered in particular the variations of standard conditions where balls were of different colours, boxes of restricted content, boxes had unequal probabilities of receiving balls, etc., but where boxes were essentially similar from the point of view of contributing to the total number of occupied boxes." Their latest paper considered "more generally the problem of sequential occupancy where the boxes are classified into different types." [9]

The Royal Statistical Society organized a meeting where J.O. Irwin read a paper titled, *William Allen Whitworth and a Hundred Years of Probability.* Given her stature and expertise in probability, David was recruited to serve as a discussant. Irwin's paper and the discussant comments were published [10, 11]. David began her discussion with the words: "I am going to strike rather a discordant note at this happy gathering...." Penn State Distinguished Professor of Statistics Donald Richards said recently: "I must say that the comments show that David indeed was tough. I imagine that Irwin himself was a little bit shaken by some of her view of his article. In any event her comments, especially those on the amazing Thomas Jarrett, are well worth reading and will always be worth reading."[31]

Thus, these times were business as usual, with David being intellectually tough and worthwhile, except for the remarkable fact that David was making the huge move to UCR. After settling in to UCR at the beginning of 1968, David wrote: "I am enjoying being here very much. Everyone is very kind and the weather is delightful—rather like a June day in England."[32] She wrote these words on UCR stationery that was printed specifically for the Biometrical Laboratory. She typed "Dept. of Biostatistics" above the pre-printed "Biometrical Laboratory."

David had many doctoral students while at UCL. Here are their acknowledgments of her assistance, found at the front of their dissertations:

1. Bateman, G.I.: "I wish to express my sincere appreciation to Dr. F.N. David for the advice and guidance given in the preparation of this thesis." (1948)
2. Mitchell, B.: "The author wishes to acknowledge the assistance and encouragement received from the Staff of the Statistics' [sic] Department, University College, London, and especially from Dr. F.N. David, under whose supervision this study was carried out." (1952)
3. Barton, D.E.: "The author would like to acknowledge his gratitude to Dr. F.N. David D.Sc. of University College, London, without whose patient and persevering criticism his task would have been more difficult, and his work done in less rewarding fields." (1953)
4. Mallows, C.: "My grateful thanks are due to my supervisors, particularly Dr. F.N. David, for their continual advice and encouragement." (1953)
5. Rudra, A.: "The investigation was undertaken under the supervision of Dr. F.N. David and Dr. N.L. Johnson to whom I express my grateful thanks." (1953)
6. Warner, B.T.: "The author wishes to express his gratitude to his supervisor, Dr. F.N. David, Reader in Statistics at University College London, for her advice and constant encouragement at all stages of the investigation." (1953)
7. Abdel-Ati, S.H.: "This research was undertaken under the supervision of Dr. F.N. David to whom the writer is indebted for advice and encouragement." (1954)
8. Bartholomew, D.J.: "My thanks are due to Dr. N.L. Johnson for his valuable advice and criticism at all stages of this work. Also to Professor E.S. Pearson and Dr. F.N. David with whom I have had profitable discussion on a number of occasions." (1955)
9. Mishra, C.: "I wish to acknowledge with gratitude and indebtedness to Professor E.S. Pearson and Dr. F.N. David for their continual kindly advice, criticism, and encouragement." (1955)

[31] Richards, D.: email to Golbeck, A.L. 17 February 2021.

[32] David, F.N.: Letter to accountant Mr. Davidson 22 January 1968.

10. Moore, P.G.: "The author would like to acknowledge the advice and suggestions that he has received from various persons during the course of the preparation of this thesis and especially from Professor E.S. Pearson, Professor J.W. Tukey and Dr. F.N. David." (1955)
11. Jenkins, G.M.: "...I am also indebted to: Dr. F.N. David for suggesting a field of study and for her interest in the work." (1955)
12. Harley, B.I.: "I would like to express my gratitude to my supervisor, Dr. F.N. David, D.Sc., for her advice and encouragement at all stages of the work." (1957)
13. Branson, M.: "The author wishes to express his gratitude to Dr. F.N. David, who supervised the investigation, for her guidance and encouragement at every stage..." (1960)
14. Crouse, C.F.: "This research was undertaken under the supervision of Dr. F.N. David, to whom the writer is highly indebted for advice and encouragement." (n.d.)
15. Merrington, M.P.M.: "I would like to thank Professor F.N. David and Professor Harry Harris most warmly for the help and encouragement they have given me as my supervisors." (1967)

Maxine Merrington, who has been called a "noteworthy statistician," was apparently David's last doctoral student at UCL. After moving to UCR, David continued to help Merrington to finish her doctoral degree. She arranged for two other faculty members give Merrington the oral examination back in the United Kingdom. Merrington was delighted at David's choice of examiners, because the two were in her words "the most agreeable examiners one could wish for." She added "and thank you for looking after my thesis so tolerantly and patiently." She continued to ask David for advice: Should she try to publish her thesis, and should she take certain measurements? David invited Merrington to come to Riverside for a visit. Merrington completed her PhD degree in 1968 and continued her career at UCL as both a lecturer in statistics and an archivist [12].

But in 1968, Merrington reported to David on some challenges in the UCL statistics department. The first year students were getting their statistics from David Hill (b.1926–d.2015), son of Sir Austin Bradford Hill who David had worked with during World War II, and their probability from Dennis Lindley: "Not surprisingly, there is a certain degree of confusion in their minds." She also reported that Lindley wanted to reduce "drastically" the amount of practical statistics that the students would learn but that she would deal with it by making sure the students very carefully wrote up the practical work that they did receive.

Also, the resources at the UCL statistics department had dwindled. Now they had no librarian, and Merrington reported from her archivist point of view that "the mess in there is indescribable. Books piled high, current journals scattered about and no sign of [the person in charge] sending anyone over ever." Also, they had only one human computer, which Merrington described as merely "better than nothing." She added "Personally I think the students are being less well looked after than previously but I may be wrong." The picture that Merrington painted for David was not a happy one.[33]

Neyman invited David to Berkeley to give a lecture in early 1968. This was part of Two Programs in Statistics: A Lecture Series on Recent Advances in Applied Statistics, and a short course on The Statistical Part of Experimental Design.[34] David's talk on Randomization Methods with Biological Applications was presented within the lecture series. Her abstract reads "A statistical approach to several problems using nonparametric techniques. The study of chromosome abnormalities, the distribution of leukemia in California, and the assessment of differences in the results of mental tests are used as illustrations. The problems of creating statistical models to solve biological problems will also be discussed."[35]

[33] Merrington, M.: Letter to David, F.N. 24 January 1968.

[34] Note: The actual title was The Statistical Part of Experimental Design X 808.

[35] University of California—Berkeley: Notice, Letters and Science Extension 24 February 1968, 9 March 1968, and 18–20 March 1968.

The cooperation between David and Neyman went in both directions. David was preparing to host someone for talks on indices of ecological diversity at the Forest Service and UCR. She made sure that Neyman had a chance to involve that person in the statistics seminars at Berkeley, and Neyman understood the person would be visiting them jointly. They worked together to arrange a schedule that included dinner and an overnight (either at David's house or the UCB Women's Faculty Club).[36,37]

Then Neyman hired David again in 1968 for three summer months to conduct research at his Berkeley Statistical Laboratory. Her title there was visiting professor and research statistician. She was funded through the US Public Health Service[38] and the State of California—Health and Welfare Agency, department of mental hygiene, Pacific State Hospital, socio-behavioral laboratory.[39] The state project was a consultancy that extended for a year until 30 September 1970 with a salary of 110 USD per day of consultation, up to a maximum of 2,000 USD. David's title at UCR at this time was professor of biometry and biological statistics.[40]

In 1968, David published another book, this time as first author with six co-authors. It was titled *Normal Centroids, Medians and Scores for Ordinal Data* [13]. As with her books on correlation and symmetric functions, this was composed of a detailed introduction (in this case to randomization tests) and how to use the tables (with examples and illustrations), followed by the many tables. The Preface stated that this book was intended to fill a gap, as there was no simple textbook for using computers and containing *necessary auxiliary tables*. The book was part of a series *Tracts for Computers,* which was once upon a time edited by K Pearson.[41]

The book contained advice concerning the practical difficulties that the authors experienced using the computer. They hoped the book would be useful to the schoolboy (referred to several times), for example, "Nowadays the schoolboy ought to be practised [sic] in computing, and this practice should run parallel with his algebraic work. He should be exercised in the use of tables which are not becoming obsolete like the smaller tables of logarithms."

The book contained three types of tables: normal centroids, normal medians, and normal scores. These were paired with auxiliary tables containing additional information (sums of squares, fourth powers, and products). Reviewer D.B. Owen (Southern Methodist University) explained: "The normal centroid is obtained by dividing a standardized normal density into $n$ equal areas and for each of these $n$ areas finding the expected value given the variable is in that particular area. Normal medians are the median values for each of these $n$ equal areas. Normal scores are the expected values of each of $n$ order statistics drawn from a standardized normal distribution."

Reviewer Owen thought the tables would be useful. But their utility would be limited by the fact that the authors gave no guidelines about which of the three tables should be used in which circumstances. He called for more research that could yield a rule of thumb [14].[42]

Reviewer H.O. Hartley (Texas A&M University) explained the circumstances where the tables in the book could be useful: "Centroids, medians, and scores, computed from the standardized normal distribution $N(0,1)$ are sometimes used in the analysis of ordinal data (i.e., ranked data) or in situations

[36] David, F.N.: Letter to Neyman, J. 26 April 1968.

[37] Neyman, J.: Letter to David, F.N. 13 May 1968.

[38] University of California: Form for change in employment status 3 June 1968.

[39] Kim, P.J.: Letter to David, F.N. 22 July 1968. Note: P. John Kim was a biostatistician. The project was titled, *Socio-Behavioral Study Center for Mental Retardation.*

[40] Williamson, R.: State of California standard agreement to David, F.N. 1 October 1969. Note: Robert Williamson was Pacific State Hospital administrator.

[41] Note: This series aimed to provide guidance with practical difficulties of using mechanical computers, and to provide useful statistical tables that were newly or already computed.

[42] Note: The Dictionary says that a rule of thumb is "A broadly accurate guide or principle, based on experience or practice rather than theory."

where it is suspected that the data on the scales on which they were originally recorded are distinctly non-normal." Hartley pointed out that the normal scores had already been tabulated, and there was already a fairly extensive treatment of the subject in the nonparametrics literature; the contribution of the book was to give more extensive tabulations. The centroids and medians on the other hand were lesser known, and their treatment in the literature more obscure.

Hartley asked some critical questions, at a time where high-speed computing was replacing tables such as the ones in this book. One question was, why the excessive accuracy in the tables [2]?[43] Another question was, why the large sample sizes when the application of normal theory was adequate at much smaller sample sizes? However, Hartley's overall assessment of the book was this: "...as a 'connoisseur of statistical tables' I raise my hat to the magnificent effort and excellent workmanship that has gone into their preparation. Moreover, users of statistical tables must undoubtedly be grateful to the enterprise for producing what is clearly the most extensive record of the integrals defining the above quantities." [15]

Despite the amazing craftsmanship that David put into her tables, the era of statistical tables in the statistics profession was reaching its end. High-speed computers were making the statistical tables obsolete.

David either never understood how E Pearson had blocked her advancement to the position of professor and head of the UCL statistics department, or else she came to accept it, which seems to us unlikely. In any case, once she got to UCR, David invited E Pearson for an extended 10-week visit to her new department, arriving at Riverside in mid-February 1969.[44] The two of them even flew up to Berkeley together to see Neyman and Scott.[45] The four of them would have dinner together and participate in the Berkeley-Stanford Colloquium on Statistics activities, and then E Pearson would give a history talk in the UCB statistics department.[46] This was E Pearson's first trip to Berkeley and Riverside.[47]

E Pearson took advantage of the time at Riverside to do some history writing.[48] As he explained, "I have been trying to put on record experiences which I went through so that later historians of statistics won't get too much wrong in ignorance. Of course I may be biased in my opinions, but on the whole I think my facts are essentially correct."[49] He went on to further explain: "My historical articles do not involve much research because I have really only written about events within my own or K. P.'s [K Pearson's] experience. Rather purposely so, because it seems a good idea to put things on record while they are in living memory, knowing how hard it is to find out information which one could have got a few years before from the living person—if, of course, he had been prepared to give it...."[50] This was different from David's approach, which involved much research about people who had been gone long enough that their work could be put into perspective.

The local Riverside newspaper described E Pearson's time at UCR as a meeting of "three old friends" (E Pearson, Neyman, and David). The article said that "All were involved in England during World War II in solving such problems as 'How many formations of planes, each carrying so many bombs of this or that capacity, are necessary to clear the mines from the landing beaches of

[43] Note: An answer appears in Barton's obituary of David, where he noted that K Pearson insisted upon 20-figure accuracy.

[44] Pearson, E.: Handwritten letter to Neyman, J. 19 January 1969.

[45] Pearson, E.: Handwritten letter to Neyman, J. 2 March 1969.

[46] Neyman, J.: Letter to Pearson, E. 4 March 1969

[47] Scott, E.L.: Letter to Knight, W.D. 11 March 1969.

[48] Pearson, E.: Handwritten letter to Neyman, J. 7 June 1969.

[49] Pearson, E.: Handwritten letter to Neyman, J. 1 December 1972.

[50] Pearson, E.: Handwritten letter to Neyman, J. 17 August 1973.

Normandy? What kind of bombs and what shape of plane formation would do the most clearing?'... 'I can remember,' he [E Pearson] said with a smile, 'that we made one U.S. manufacturer happy because we found that in certain circumstances a certain type of small bomb, dropped by parachute in large numbers, would be most effective.'"

Pearson was called a "titan," and Neyman and David were called "two old and prominent professional associates with whom he had worked during his 25 years as chairman of the Statistics Department at the University of London." The article went on to say that "The three friends [E Pearson, Neyman, and David] could give virtually an eye-witness view of the whole development of statistics' present place in science."[51] The academic visit in 1969 must have gone well, because E Pearson began planning another visit to UCR and UCB the next year.[52]

Now that both David and Neyman were in California, they could meet and collaborate more frequently. In addition to her summer work at UCB, David sometimes hosted Neyman to speak at UCR. For example, on 11 April 1969, Neyman traveled to UCR to attend a seminar and give a lecture on "Statistical studies of certain chance mechanisms in developing populations and in their competition." UCR paid for his transportation and car expenses.[53]

Merrington in the United Kingdom stayed in touch with her mentor David in the United States. Merrington indicated that her new department head, Lindley, wasn't very attentive of her needs. She was discouraged: "I pity the [students] allotted to me. I told D.V. Lindley I wasn't competent but he didn't take it in. I would like to retire."[54]

As department chair at UCR, David actively advocated for additional computer resources. She wrote to her vice-chancellor and dean: "Whether we like it or not the computer age in scientific teaching and research is with us and I have been giving some time and thought lately to the organization of this department to try to meat [sic] the projected computer needs of the college over the next few years. The growth of the use of computers is such that any projections are likely to be underestimates." David then laid out the funds needed and the funds allocated, as well as projected needs for computer instruction and usage and for data reduction and analysis.

The letter was highly detailed and showed that David had done her homework. She concluded the letter by saying "It is difficult in this day and age to look ahead. I have no doubt that the present fashion for computer usage will continue beyond the time I have left in the university world. I think, as the campus grows... the need for statistical advice and the need to prevent the numerically illiterate wasting computer time will grow even faster. We have to cut our coat according to the size of the cloth but... I am hoping that eventually on campus there will be a department of Numerical Science which will relieve us of formal teaching of computer usage, but we will want the Programmers then to help in the research field."

David was a good leader and visionary, as indicated in the conclusion to her letter: "The new faculty will have 11-month appointments and will spend two months each year on statistical research... I would hope that they will also consider it part of their obligation to the university—as I consider it part of mine—to help any colleague or student who asks for help... In recruiting new faculty I am keeping in mind the idea that such an interest may be obligatory for any statistician in the nineteen-eighties."[55]

---

[51] Patterson, T.: World-famous statisticians renew wartime ties during UCR lectures. Riverside, CA: Press-Enterprise B-4, Saturday, 19 April 1969.

[52] Pearson, E.: Handwritten letter to Neyman, J. 9 March 1970.

[53] Neyman, J.: Document on travel 1969.

[54] Merrington, M.: Handwritten letter to David, F.N. 29 January 1969.

[55] David, F.N.: Memorandum to Jenkin, Dugger, and Day 26 May 1969. Note: Jenkin was vice-chancellor, Dugger was dean, and Day was associate director.

Based on our own experience at the time, California was a place of high political awareness, and Neyman was interested in public policy. Thus, we are not surprised that David, by association, became involved in state politics. For example, she expressed interest in, and supported creation of, an advisory board to the state legislature that had to do with science. The Speaker of the California Assembly informed her of its creation: "You may be pleased to know that the California State Assembly has acted favorably on the resolution formally establishing the Assembly Science and Technology Advisory Council (H.R. 190)." He articulated that there was widespread interest in a science council, and they would act quickly to find the right chairperson for the council. He welcomed David's continuing thoughts on the membership or structure of the council and made assurances that her recommendations would receive careful consideration.[56]

O'Regan, David's primary contact at the Forest Service, wrote this evaluation of her work: "The data indicate that Dr. David is a productive research worker. I know from personal experience that she is an outstanding teacher and an excellent consultant. Her teaching abilities and experience range from incoming freshmen to post doctorals and from theoretical to applied courses. She is particularly efficient and creative in handling data."[57] It was a succinct explanation of David's breadth of expertise.

## 8.3 Always a Researcher

By October 1969, the Griffin publishing company was putting out a paperback version of David's *Games, Gods and Gambling* book. They used a new design, and the cost of the book was to be 25 GBP. The book was to be published in the United States.[58]

At the same time, David was negotiating with Griffin for a new edition of her 1953 statistical primer book. She and the publisher were having some disagreement. The publisher tried to placate her:

> I am most disturbed that you should have considered any comments of mine on your proposals for augmenting the [A Statistical Primer] book damping or lukewarm. In fact, both Alan Stuart and I were very pleased that the treatment you felt best for the book would be the addition, at the end, of chapters on correlation and regression and that, as far as we can gather, little else would have to be done to the text. [para] If in my letter of 9th July I failed to convey this I am very sorry because no such impression was intended. The acceptance of the additional chapters was implied, perhaps less pointedly than it should have been....

The publisher was looking for a revision by the end of November 1969 with a title that was different from the original. The new title would be *A First Course in Statistics*.[59]

In 1969, Christopher A. Robertson (b.1942–d.2019) was a lecturer in mathematical statistics working on his dissertation at the University of Exeter. In the course of receiving mentoring from E Pearson, both in writing and in person, Robertson expressed admiration for David's work: "For the ML estimators (which I have not yet looked at except superficially) the second-order calculations must be rather alarming. Professor David must have had considerable trouble in obtaining her results. I think I can see how one could carry the job through, at any rate in principle, but the details look formidable."[60]

56 Monagan, B.: Letter to David, F.N. 10 July 1969. Note: Bob Monagan was speaker of the assembly.

57 O'Regan, W.G.: Letter to David, F.N. 25 July 1969.

58 Griffin, R.: Letter to David, F.N. 22 October 1969. Note: Rae Griffin was an executive at Charles Griffin Publishing Company Limited.

59 Griffin Publishing Company Executive: Letter to David, F.N. 22 October 1969.

60 Robertson, C.A.: Letter to Pearson, E. 22 October 1969.

E Pearson reported to David on Robertson's research progress, in considerable statistical detail, which indicated E Pearson and David continued to have an intellectual relationship after she moved to California. David thought highly of Robertson, and their intellectual interests overlapped. She asked E Pearson to find out if Robertson had any interest in visiting UCR. E Pearson reported that Robertson was "quite appealed" with the idea.[61] The next year, David brought him to UCR as a visiting professor. He moved to a regular faculty position and stayed at UCR for 32 years, earning tenure and making many contributions to the department and profession [16].

England passed the 1965 Rent Act. It brought in "new rights for tenants and landlords and a new way of helping them to agree on fair rents."[62] The Act specified protections against eviction, protections for unfurnished lettings and protections for furnished lettings. It was followed by the 1968 Rent Act, which made further specifications regarding fair rent.[63] David paid attention to these acts because, after she moved to California, she kept occupancy of the 28 Torrington Square apartment. By English law, she was a regulated tenant of that property.

UCL did not want to increase the rent at 28 Torrington Square, at least not in 1969,[64,65] but they did want to register for fair rent. Under the Fair Rent Act, David was given the opportunity to dispute the prescribed rent.[66]

The university wanted to start charging for cleaning. David disputed this new charge under the Fair Rent Act: "I wish to ask for a reduction of the sum charged for services (...) on the grounds that the amount of traffic on the stairs is so small that the tenants of N. 28 think cleaning once a fortnight, instead of once a week, would be adequate. [para] I should like to add that we were involved in the expense of cleaning and maintaining this carpet without being asked if we wanted it. I personally did not, and we have already given my views to Messrs Clutton."[67]

While David was in California, there were two people living in her two-bedroom apartment at 28 Torrington Square. One was an unidentified person, and the other a person named Helen. David registered the dispute after receiving a letter from the unidentified person, who told her: "I hate putting anything extra on to you when I know you have more than you can cope with now. But I feel you ought to scrutinize my letter [to the landlord]." The person outlined the rent issues and said they would have moved out if the rent were raised: "If necessary I'll find a room in someone else's house! (Helen will definitely have to give it up if her rent goes up.)" The unidentified person was definitely close to David. The letter began with: "Darling David." It ended with: "Very much love."[68] The letter was signed "Jo." From a subsequent letter, we recognized that Jo was Joyce Townend.[69]

A move across continents can be disrupting for any person. So it was for David. For example, she needed to make sure that statistics journals like the *Journal of Applied Probability* and *Advances in Applied Probability* were being sent to her in California.[70]

---

[61] Pearson, E.S.: Letter to David, F.N. 5 November 1969.

[62] Ministry of Housing and Local Government and the Central Office of Information: The 1965 Rent Act, Printed in England by Her Majesty's Stationery Office 1965.

[63] Ministry of Housing and Local Government and the Central Office of Information: The 1968 Rent Act (Prices and Incomes Act) 1968.

[64] Cluttons: Memorandum to David, F.N. 24 October 1969.

[65] Cluttons: Memorandum to [David, F.N.] 28 October 1969.

[66] Bootman, B.: Notice to David, F.N. 4 November 1969. NOTE: B. Bootman was a rent officer.

[67] David, F.N.: Letter to rent officer, Swinton House October 1969.

[68] Jo: Letter to David, F.N. 4 November 1969.

[69] Townend, J.: Handwritten letter to David, F.N. 26 August 1971.

[70] David, F.N.: Letter to Hitchock, M. 6 January 1971. Note: M. Hitchcock was Journal of Applied Probability editorial assistant.

Another example of an impediment is that David left notes in London, and that made it difficult for her to respond to letters. In a reply to Olar Reiersol in the Matematisk Institutt at the University of Oslo, she wrote "This is an answer to your letter of February 8, 1971; and, since I am far from my history notes and papers briefly the answer is I do not remember... This is all I meant or so I think after a gap of ten years or so from my writing... I find J.F. Scott *A history of Mathematics* as an accurate writer. I forget what he says now and my copy of his book is in England with the rest of the notes I made at the time... I have been here now for three years (plus) starting a new department of Statistics and having little time for reflection and history musings. Perhaps when there is more money in the University of California than there is now you might care to pay us a visit."[71]

David was highly interested in the sport of cricket. This is a sport first played in southeast England in the sixteenth century, so we weren't surprised to discover that David was a chief patron of the UCR Cricket Club. When her club applied to become an official member of the Southern California Cricket Association, they made sure to let David know, and they pledged that they would let her know the outcome of their application.[72]

Cambridge University Press was considering the future of David's book, *Probability Theory for Statistical Methods.* The book was first published in 1949, and by 1971 it was only selling at about 100 copies a year. The supply was running low, and Cambridge proposed to David that they let the book go out of print in about a year, by which time the rest of the copies would have been sold.

The publisher asked: Did David want any of the copies for herself that were still available? They also expressed some encouragement about the work: "This is not necessarily the end of the story. The very fact that a book is out of print sometimes stimulates demand to a level where it is possible for us to resuscitate it ourselves in some form. In any case, when any of our books has to be declared out of print in our own edition, we automatically try to interest another publisher in the possibility of reissuing it."[73]

David was glad to hear about the remaining copies of her book and said she would order some from their New York office. She told her publisher she hadn't thought about reprinting the book but that "there may be a case for a drastically revised version"[74] The field had changed in the 22 years since the book was published, and David knew it.

Shortly thereafter, the McGraw-Hill Book Company got in touch with David. They were interested in reviewing a manuscript from her, but they didn't have one to review, nor did they have a history of her developing materials for a book with them. They wanted to make things right with her and wanted to hear from her.[75]

David's administrative work sometimes got in the way of her research and consulting work. For example, she declined an invitation to chair a hydrology session at a professional meeting, and she communicated a delay in getting some papers finished, all due to time constraints: "I can't be sure that I will have time to do anything worth while. [para] Sorry to be so tiresome but things are pretty silly here in that the administrative work seems never ending."[76]

---

[71] David, F.N.: Letter to Reiersol, O. 17 February 1971. Note: Olar Reiersol was University of Oslo Matematisk Institutt professor.

[72] Khasimuddin, S.: Letter to David, F.N. 14 January 1971. Note: Syed Khasimuddin was University of California - Riverside cricket club captain.

[73] Unidentified: Letter to David, F.N. 19 March 1971. Note: The letter was apparently from someone at Cambridge University Press.

[74] David, F.N.: Letter to Eccleshare, C.F. 1 April 1971. Note: C.F. Eccleshare was Cambridge University Press employee.

[75] Aksen, H.S.: Letter to David, F.N. 13 April 1971. Note: Howard S. Aksen was McGraw-Hill Book Company senior editor.

[76] David, F.N.: Letter to Moran, P.A.P. 26 April 1971. Note: P.A.P. Moran was Australia National University Institute of Advanced Study statistician.

In 1970, President Nixon set up a Commission on Federal Statistics to study the Federal statistical system: how it operates, how it responds to users, and how it works for people who supply data. They contacted David because she had "useful ideas about the improvement of statistics in [her] agency," and they asked her to complete a questionnaire.[77]

In 1971, David started a technical report series at UCR. It included, among other things, lectures by E Pearson on the history of biometry.[78] E Pearson went on to publish (Charles Griffin and Co.) an edited book, with Maurice Kendall, on *Studies in the history of statistics and Probability.*[79]

David was living in the University Village Apartments.[80] These are located adjacent to the UCR campus, in fact they are the closest apartments to the campus. Thus, David was able to enjoy the close proximity to her campus in a similar way that she was able to enjoy the close proximity to UCL from 28 Torrington Square.

Neyman continued to promote David in the research world. The National Academy of Sciences Council was formulating several research projects, one of which was connected to the Environmental Studies Board and had to do with pest control. Neyman was especially interested in this project and connected David to it: "I wish to assure you that the success of this study depends very much on vision, initiative and competence of the statistician whose services will be unavoidable. My immediate purpose is to recommend to your attention Miss F.N. David, Chairman of the Department of Statistics, University of California, Riverside, 92502. I have known Dr. David for many years now, have a great respect for her work conducted on a variety of subjects and am sure that her participation in the pest control study will be most beneficial." He told them that David was currently already involved in a similar study for the National Science Foundation.[81] In turn, the academy expressed interest in David's expertise.[82]

David then described the current project, which would explore pests on cotton, alfalfa, citris, and other plants. The idea was to conduct a "cost/benefit study of the strategies involved in biological control or insecticides or both." There would be a consortium of agricultural institutions participating, and David's department at UCR would be the coordinating center for the analysis. She had seed money from UCR to carry out a literature search and review data from previous non-published studies. She explained "This will enable us to see what further experiments need to be performed and what further investigations will possibly be required, before the main project begins." The National Science Foundation had already committed 100,000 USD to plan the project: "The U.C.R. Departments of Entomology and Statistics have been ready with their outline of work for some months but since the proposal is country-wide it takes time to get everyone in harmony."[83]

---

[77] Rathbun, D.B.: Letter to David, F.N. 18 April 1971. Note: Daniel B. Rathbun was President's Commission on Federal Statistics executive director.

[78] David, F.N.: Letter to Dryer, P.S. 15 July 1971. Note: Paul S. Dryer was at the University of Michigan department of statistics.

[79] Plackett, R.L.: Letter to David, F.N. 18 April 1974. Note: R.L. Plackett was at The University of Newcastle Upon Tyne department of statistics.

[80] Kaplan, J.M.: Memorandum to David, F.N. n.d. Note: Jay M. Kaplan was University Village Apartments' secretary-treasurer.

[81] Neyman, J.: Letter to Handler, P. 13 August 1971. Note: Philip Handler was National Academy of Sciences president.

[82] Zucker, A.: Letter to Neyman, J. 15 September 1971. Note: Alexander Zucker was National Academic of Sciences Environmental Studies Board executive director.

[83] David, F.N.: Letter to Neyman, J. 13 August 1971.

Joyce Townend wasn't just living at 28 Torrington Square. She was also using David's cottage at 5 Stuart's Meadow, Graffham, Nr. Petworth, Sussex, England. Graffham is one of the prettiest villages in England. We can imagine that David retreated to her cottage to get some peace, walk, and perhaps remind herself of days growing up in the countryside.

David and E Pearson continued to have a relationship. He was not in the best of health in 1976, but he expressed hope that he would be driving a car again soon and would go to David's bungalow in Graffham. "I have passed through Graffham a number of times but never have been able to identify your bungalow, which shows a shocking lack of memory as I stayed there with you at least once some years ago," he wrote. He also inquired whether David was going to be in Riverside for another year or if she would be at her house in Berkeley.[84]

Joyce was planning to go to the Graffham cottage for about a month. She was looking forward to meeting David at Heathrow Airport and then spending a "Very Quiet Weekend" with her. Again signing her name "Jo," Joyce addressed her letter to "Darling David."[85]

We think David at this point was keeping four residences: the flat at 28 Torrington Square in London, the cottage in Graffham, the house in Kensington Berkeley, and the apartment in Riverside. One of David's neighbors, probably in Berkeley, was watching over her property there and gave David a report. One of the trees had died, the trash cans were full, and one of the neighbors was having a problem with the police. David was away on vacation. We surmised she was in London visiting her elderly mother.[86]

Also at this point in 1970, David was teaching on two university campuses. She was teaching within her professor and chair role at UCR, but she also began teaching a course at UCB. This was her History of statistics course, and she taught it regularly at Berkeley for a number of years, quickly attracting over 450 students [17][87]. The course at UCB eventually became too popular to keep under control, and so she gave it up: "When you become a sort of favorite sport of the football fans, you know that you're doing something wrong," she said [3].

David was extremely busy, but she was still willing to carve out time to help a student. For example, J.K. Ghosh in Calcutta at the Indian Statistical Institute wrote to her asking if she would read a student thesis, and she agreed.[88]

David also continued to be committed to her profession. She wrote "careful and thoughtful;" "thorough, concise, and specific;" and "helpful" reviews of papers for journals, such as *The American Statistician*.[89] When she did her reviews, she asked herself two questions: Is it of practical use? and Is it a genuine mathematical contribution?[90]

There wasn't time to do everything.[91] G.P. Patil in the statistics department at Penn State University invited her to visit in the summer of 1972, but she had already committed to teaching in the summer session at UCB. She hoped Patil would ask her again sometime.[92] Neyman invited David to contribute

---

[84] Pearson, E.S.: Letter to David, F.N. 9 August 1976.

[85] Townend, J.: Letter to David, F.N. 26 August 1971.

[86] Rose [last name not indicated]: Handwritten letter to David, F.N. 17 September 1971.

[87] Author not named: David, F.N. 23 August 1909, Journal not named, n.d.

[88] David, F.N.: Letter to Ghosh, J.K. 28 [month not indicated] 1971. Note: J.K. Ghosh was Indian Statistical Institute PhD committee convener.

[89] Gibbons, J.D.: Letters to David, F.N. 29 June 1972. Note: Jean D. Gibbons was *The American Statistician* associate editor.

[90] David, F.N.: Letter to Shenton, L.R. 7 January 1974. Note: L.R. Shenton was at the University of Georgia and was *Journal of the American Statistical Association* editor.

[91] Kendall, M.G.: Letter to David, F.N. 7 July 1972. Note: Letters went in both directions. Also Note: M.G. Kendall was Scicon editor.

[92] David, F.N.: Letter to Patil, G.P. n.d. Note: G.P. Patil was at the Penn State University statistics department.

to his book on Nicolaus Copernicus that would be themed "Indeterminism in Science." She gave him a list of ideas for her possible contribution:

> Statistics: Making common sense of the data of experience.
>
> ...
>
> In controlled experiments there is the opportunity for deliberately introducing randomness into the design so that systematic variation can be separated from purely random error. Weakness of experimental designs under certain conditions.

David planned to write about the work of Galton, Pearson, Gosset, Fisher, and Neyman.[93] After all, she had known each of them personally. But time escaped her, and she ended up withdrawing from the Copernican Volume.[94]

David was usually game for helping out a colleague or her university. For example, she advised colleagues at the University of California—Los Angeles on developing an interdisciplinary undergraduate program in human biology. Her contact at UCLA wrote "I carried away from our discussions the impression that, with a concerted effort (and much hard work!), it should be feasible to develop a quite exciting curriculum with strong undergraduate appeal." Statistics curriculum development was an area where David had developed "deep insight" into its many factors.[95] Furthermore, she served on a UC president's study committee that included members from across the system.[96] For another example, David agreed to serve on a panel about Affirmative Action for the American Society of Public Administration, where the broad theme was "New Developments in Affirmative Action—City, County, and University." She was asked to give some data on the problem in the university, along with some suggestions for positive change.[97]

Later, David was asked as a personal favor to help adjudicate a dissertation at a different university.[98] David reluctantly agreed and included a warning that she should only be used "if you're sure I'm the right person. I have a reputation for being rather harsh about these things. [para] You will note I didn't say thank you for your letter."[99]

## References

1. Beaver, R.J. and Arnold, B.C.: In Memoriam: Morris J. Garber. (2022) https://senate.universityofcalifornia.edu/_files/inmemoriam/html/morrisjgarber.htm Accessed 4 October 2022
2. Barton, D.E.: Obituary: Florence Nightingale David. Journal of the Royal Statistical Society, Series A (General) 157(2), 299–303 (1994)
3. Laird, N.M.: A conversation with F.N. David. Statistical Science 4(3), 235–246 (1989)
4. Hand, D.J.: Dennis Lindley obituary: Mathematician who was a leading figure in the Bayesian school of statistics. The Guardian 16 March (2014) https//www.theguardian.com/science/2014/mar/16/dennis-lindley Accessed 15 June 2022
5. Royal Statistical Society: Interview with Dennis Lindley (2022) https://www.youtube.com/watch?v=cgclGi8yEu4t=539s Accessed 15 June 2022

---

[93] David, F.N.: Letter to Neyman, J. 27 December 1971.

[94] David, F.N.: Letter to Neyman, J. 12 April 1972.

[95] Adey, W.R.: Letter to David, F.N. 6 December 1972. Note: W. Ross Adey was UCLA anatomy and physiology professor.

[96] Taylor, A.E.: Letter to David, F.N. 1 December 1972. Note: Angus E. Taylor was University of California vice president—academic affairs.

[97] Jebens, A.B.: Letter to David, F.N. 5 December 1972. Note: Arthur B. Jebens was University of California—Riverside vice-chancellor—business and finance.

[98] Marr, A.G.: Letter to David, F.N. 11 April 1974. Note: Allen G. Marr was University of California - Davis dean.

[99] David, F.N.: Letter to Marr, A.G. 18 April 1974.

6. Leonard, T.: The life of a Bayesian boy, Chapter 3: University College London (2022) http://thomashoskynsleonardblog.blogspot.com/2021/04/the-life-of-bayesian-boy-tom-leonards.html Accessed 16 June 2022
7. Barton, D.E.: Obituary: Florence Nightingale David, Journal of the Royal Statistical Society Series A 157(2) 299–303 (1994)
8. Godfrey, A: The Search for Forest Facts: A History of the Pacific Southwest Forest and Range Experiment Station, 1926–2000, United States Department of Agriculture General Technical Report PSW-GTR-233 April (2013) https://www.fs.usda.gov/psw/publications/documents/psw_gtr233/psw_gtr233.pdf Accessed 23 June 2024
9. Barton D.E. and David, F.N.: Sequential occupancy with classification. Biometrika 55(1), 229–241 (1968)
10. Irwin, J.O.: William Allen Whitworth and a hundred years of probability. Journal of the Royal Statistical Society, Series A (General) 130(2), 147–176 January (1967)
11. David, F.N.: Discussion on Dr. Irwin's paper. Journal of the Royal Statistical Society, Series A (General) 130(2), 165–168 March (1967)
12. Prabook, M.P.: Mary Merrington (2022) https://prabook.com/web/maxine_patricia_mary.merrington/365131 Accessed 14 October 2022
13. David, F.N., Barton, D.E., Ganeshalingam, S., Harter, H.L., Kim, P.J., Merrington, M. and Walley, D.: Normal Centroids, Medians and Scores for Ordinal Data. Cambridge University Press (1968)
14. Owen, D.B.: Review of David, F.N., Barton, D.E., Ganeshalingam, S., Harter, H.L., Kim, P.J., Merrington, M. and Walley, D. Normal Centroids, Medians and Scores for Ordinal Data. Journal of the American Statistical Association 64(328), 1684–1686 December (1969)
15. Hartley, H.O.: Review of David, F.N., Barton, D.E., Ganeshalingam, S., Harter, H.L., Kim, P.J., Merrington, M. and Walley, D. Normal Centroids, Medians and Scores for Ordinal Data. Review of the International Statistical Institute 39(1), 197 (1971)
16. University of California - Riverside: In Memory of Christopher Robertson, College of Natural and Agricultural Sciences (2020) https://cnas.ucr.edu/news/2020/01/03/memory-christopher-robertson. Accessed 25 July 2022
17. Lehmann, E.L.: Reminiscences of a Statistician: The Company I Kept. Springer, pp. 116–119 (2008)

# Leading (1973–1977)

# 9

*In addition to [her] purely scholarly activity, FND [Florence Nightingale David] exhibited an enviable organizational talent. What I hear about her Department at Riverside is very impressive.*

Neyman, J.: letter of recommendation for David, F.N. (1972)

## 9.1 Growing a Department

The UCR department of biostatistics was approved in January 1968 and first appeared in the 1968–1969 UCR academic catalog. David was listed as chair. UCR was small at the time (approximately 5,000 students).[1] David's annual salary was 21,800 USD, and she received a 450 USD stipend for being department chair for the last 5 months of the fiscal year.[2] The new biostatistics department was composed of four people, two senior and two junior. David and Morris J. Garber were professors of biostatistics. Thomas M. Little (biometry) and John K. James (computer methodology) were lecturers.

David was listed in the catalog as being responsible for seven courses. Three were at the undergraduate level: a two-semester sequence in Introduction to statistics and nonparametric techniques. An additional four courses attached to David's name were at the graduate level: Nonparametric methods, Statistical ecology, Seminar in statistical ecology, and Seminar in problems of health and biology. In fact, these four courses were all of the graduate-level courses that were offered by the department.[3]

In 1969–1970, two assistant professors in residence and an additional lecturer (biostatistics) were added to the UCR biostatistics faculty. With additional faculty members, David was able to assign courses to others and reduce her teaching responsibility from seven to four courses. The catalog announced that the department was concerned with statistical consulting in addition to the teaching of statistics: "The courses offered present a comprehensive spectrum of statistical and probability theory in so far as such theory is necessary for the understanding and analysis of observational data." There were lecture courses, lab courses, and seminars. A student could earn either an AB or a BS

[1] David, F.N.: Letter to Patil, G.P. 14 August 1973. Note: G.P. Patil was in the Penn State University statistics department.

[2] Form for Change in Employment Status, University of California 3 June 1968.

[3] University of California Riverside Academic Catalog 1968–1969.



A. L. Golbeck, *Florence Nightingale David*,
https://doi.org/10.1007/978-3-031-77768-4_9

degree in biostatistics. Following in the UCL and UCB traditions, majoring students were required to have a solid foundation of mathematics.[4]

In 1970, the name of the department was changed from biostatistics to statistics, and David's title became professor and chair of statistics. The major was still in biostatistics, rather than being changed to statistics. There were now two professors (David and Garber, as before), two assistant professors, and five individuals at lower ranks. David's teaching load and courses remained as they were.[5]

In 1971, the degree program names were changed from biostatistics to statistics, and David added her History of probability course (Games, Gods and Gambling) to the curriculum. Two more assistant professors were added to the department, for a total of four regular (not acting) assistant professors.[6]

Thereafter, David made it clear that she was leading a statistics department and not a biometry department. She said "...the only Biometry we do is in the form of advice to members of the Agricultural Experiment Station." Her attitude was: Please don't send me any biometry books to review for possible classroom use.[7]

In 1972, a PhD degree in applied statistics was added, as well as an interdepartmental PhD degree in population biology. The faculty continued to grow. Three more regular (not acting) assistant professors were added, for a total of seven regular assistant professors. With one less faculty member at lower ranks, there were now 13 (including David) faculty members in the department. The statistics department, together with the departments of biology and entomology, began offering an undergraduate major in systems ecology. In each of the next 4 years, from 1973 to 1977, the number of faculty members in the department was either 15 or 16.[8,9,10,11,12]

In summary about faculty, David built a frequentist biostatistics department and rebranded it as a statistics department. She quadrupled the size of her faculty, growing the department from 4 faculty members to 15 or 16 in just a few years. Following the model of the UK universities, she built a bottom-heavy department: At the top, there were only two full professors (herself and Garber who was at UCR before she was hired); the rest were junior faculty members (mainly assistant professors and instructors). The custom in David's time in the United Kingdom was that there was one professor, and then there were people who were junior to that person. When she went to UCR, apparently, that was her vision of how to construct a department. As a result, David was perceived as doing a lot of things at UCR that others would have considered overload work. If she had hired some more senior people, she could have delegated such work.

In summary about curriculum, David oversaw the conversion of bachelor's degree programs in biostatistics into statistics programs. Later she partnered with two other departments to build a bachelor's degree program in systems ecology. She also built a PhD program in applied statistics [1].[13] Along the way, David taught her beloved subject, the History of probability and statistics (Games, Gods and Gambling).

---

[4] University of California Riverside Academic Catalog 1969–1970.

[5] University of California Riverside Academic Catalog 1970–1971.

[6] University of California Riverside Academic Catalog 1970–1971.

[7] David, F.N.: Letter to Miller, H.C. 2 January 1974. Note: H. Clinton Miller III was in the Medical University of South Carolina department of biometry.

[8] University of California Riverside Academic Catalog 1972–1973.

[9] University of California Riverside Academic Catalog 1973–1974.

[10] University of California Riverside Academic Catalog 1974–1975.

[11] University of California Riverside Academic Catalog 1975–1976.

[12] University of California Riverside Academic Catalog 1976–1977.

[13] Note: Nancy Sue Johnson, under David's direction, was the first PhD to graduate from the department.

In 1971, David described the situation as "pretty hectic." She became weary, telling E Pearson "I try to do too much and only succeed in doing nothing properly. The U. of C. is in a changing state—not enough money, therefore change in graduate versus undergraduate admissions—money out for agriculture, therefore curtailment of programs—and so on and so on. I get a bit tired of it all because there is lipservice to democracy and so everyone is allowed their say, but I see enough behind the scenes to realize that the decisions have all been taken before the pseudo-democracy gets to work. I think the U. of C. has asked for all it is getting but it is an anxious time for us in that I don't care for myself but have to try to protect my young faculty. So far I have succeeded in not being cut anywhere, but what will happen in 1972/3 I can't forecast. The saving grace for us is that student numbers in the statistics courses are rising... If we can only teach well enough for the numbers to go on rising in 1971/2 we may escape the axe once again... [para] I am a lousy correspondent, not because I don't want to communicate, but because time for writing seems hard to come by."[14]

Also in 1972, David was spending a good amount of time trying to secure funds for computing. She was abroad when she learned that her dean at UCR had given her additional computer use funds.[15,16] Specifically, he gave her 4,000 USD, which apparently was at least part of what she was promised. The dean's office asked her to use the funds in the current fiscal year, so that they could demonstrate the need for the next fiscal year. Except for this good news, David said she was sorry she "ever came back" to Riverside.[17]

The letter indicated that David had a good relationship with her dean. His office told her that they missed her, hoped she was enjoying her leave, and hoped she wasn't working too hard. They sent a "hi" from the dean and said he wished her a good time. Finally, they reassured her: "... we are looking after things for you."[18]

David's reputation as a department chair was growing. A person with a doctorate in mathematics wanted to go to UCR to earn a master's degree in statistics. David explained that UCR had no master's degree in statistics (at the time), but there were still advanced opportunities that she could offer him. He would be welcome to come and spend a year in the department taking statistics courses. She could get him a visiting faculty position, which would enable him to attend courses for free. She also told him that she might also be able to get him a part time teaching position.[19]

The head of the computer science and statistics department at California Polytechnic State College (Cal Poly) wrote this to David: "Mr. Robert Butler is full of praises for your new program. He has demonstrated by what he has shared with us how well you educate your students." Butler was finishing his PhD degree in statistics at UCR, and he was going to work at Cal Poly. "We are pleased that [Butler] has received such a strong background and will use it to contribute in our program," the chair wrote.[20] The statistics major was just becoming established at Cal Poly, with Bob Butler and the other

---

[14] David, F.N.: Handwritten letter to Pearson, E.S. 9 April 1971.

[15] David, F.N.: Letter to Ranwell, C.K. 4 December 1972. Note: C.K. Ranwell was secretary at Charles Griffin and Co, Ltd.

[16] Note: At this point in time, David was not earning any salary from the United Kingdom, except for book royalties from Charles Griffin publishers. Therefore, the new UK Value Added Tax did not apply to her.

[17] David, F.N.: Letter from to Gerald, C.F. 30 March 1972. Note: Curtis F. Gerald was the head of the computer science and statistics department at California State Polytechnic College.

[18] Skotty, P.: Letter to David, F.N. 31 January 1972. Note: Pat Skotty was a worker in the office of the dean, College of Biological and Agricultural Sciences, University of California—Riverside.

[19] David, F.N.: Letter to Stodghill, J.R. 30 October 1973. Note: Jack R. Stodghill was in the Dickinson College department of mathematics.

[20] Gerald, C.F.: Letter to David, F.N. 30 March 1972. Note: Curtis F. Gerald was head of the computer science and statistics department at California Polytechnic State College, now California Polytechnic State University.

statisticians and computer scientists leaving the mathematics department and establishing their own department in 1970 [2].

This was an opportune, yet stressful, time to be an administrator of a statistics department in the United States. The field of statistics was growing and expanding. Academics across the spectrum of the sciences—and especially mathematics—were wanting part of the action…and the students. The academic politics were thick. For example, the mathematics department at UCR managed to delay approval of some biostatistics courses.[21] David said that she "got in the middle of a battle in Riverside which really was fantastic. I had no idea things like this went on…the mathematicians tried to swallow me. I fought them on the senate floor and all over the place. And various other departments thought it would be a good idea. You know I ended up, this always made people laugh, I ended up a Professor of Statistics, I was a Professor of Economics, in all I had 5 different titles. They were all trying to get a statistical monopoly." [3]

In 1972, UCR's vice-chancellor for academic affairs called together a thought group to plan the future for statistics at the university. He invited four faculty members to an ad hoc committee. Among them were David and the chair of political science. The aim was to determine whether the campus was doing the important task of teaching statistics in the most efficient and effective manner. The vice-chancellor wrote "Apparently many departments give and need specialized statistics courses, but it is also evident that a basic statistics course for all students could be devised. Unavoidably such a course would have many consequences in terms of workload as well as relationships between departments and colleges. It is very desirable at this time that the matter be studied and the possibility of other options be considered." He asked faculty member J. Carlsen to chair the committee, rather than asking David who was leading the statistics department.[22]

All across the country, there was angst about who should be teaching statistics in the university. The head of the statistics department at Kansas State University decided to collect information about what was being done at the other universities. While David's responses have not survived, the letter soliciting her responses is informative about the climate that David was administering in:

> A committee at Kansas State is making a study of the diffusion of instruction in probability and statistics which has come about here over the past 10–12 years. I am not on the committee but have an obvious interest in ascertaining both the true climate here and at other schools with a department of statistics.
> It has been my belief from the time I came here, and also that of my statistical colleagues as they joined me, that all basic theory and methods should be taught by a Department of Statistics. It was fine with us if other disciplines such as psychology, education, economics, business, and engineering taught specialized courses based on one or more prerequisites from our department. At the time the Department of Statistics was created in 1959, and for some unknown time thereafter, the situation conformed to our policy. Then certain departments gradually began to teach even the most elementary and basic probability and statistics under the guise of teaching applications; and, usually in a graduate-credit course. This naturally has three harmful consequences, from our point of view:
>
> 1. It cuts down our enrollments; and, enrollments are the basis for the granting of funds, to a considerable degree;
> 2. The basic materials being taught by these departments are inferior in quality and quantity even though being taught in graduate courses in some instances; and,
> 3. At Kansas State the amount of credit you get for your efforts depends not only on the number of enrollees but also on the level of the course; hence these academic poachers receive from two to four times the credit for their teaching of elementary statistics as we do because they include it in graduate courses!
>
> I am wondering what the situation is at your institution with regard to the matters described above.[23]

[21] Patterson, T.: World-famous statisticians renew wartime ties during UCR lectures. Riverside, CA: Press-Enterprise B-4, Saturday, 19 April 1969.

[22] Golino, C.L.: Memorandum to Carlsen, J., Carter, L., David, F.N., and Reagan, M.13 April 1972. Note: Carlo L. Golino was University of California—Riverside vice-chancellor—academic affairs.

[23] Fryer, H.C.: Letter to David, F.N. 7 August 1973. Note: H.C. Fryer was head and director of the Kansas State University department of statistics.

This letter resonated with us, as the author of this book was head of a statistics unit at the time and ruminated about this type of situation. David was head of a statistics department when, given the external pressures and attempts at poaching, it wasn't an easy time to be such an administrator.

In spring 1974, the UCR department of statistics was scheduled to move into a new building, along with the departments of biochemistry, nematology, plant pathology, toxicology, and physiology and the computing center. Statistics would share the second floor with plant pathology. Celebrate! There would be various jamborees in May and June.

One event was an "Open House" on May 31. Each department in the new building would have a distinguished speaker. David invited Neyman to be the distinguished speaker who would "give a popular exposition" of the subject of statistics.[24] Neyman chose as his theme "The History of Statistics in the USA." In addition to the lectures, there were exhibitions and demonstrations. For statistics, there were:

> Demonstrations on the new Data Transmission Terminal, the APL terminals at any time. Operational demonstration of the Hewlett-Packard plotting equipment. Exhibits consist of diagrams and charts along the corridors. There are special hand machines in the laboratories.[25]

David eagerly began inviting people to attend[26,27] while she waited for Neyman's positive response, which he promptly delivered.[28] Neyman and Elizabeth L. Scott flew down together from Berkeley and attended the celebration.[29]

David continued to work on developing a statistics library at UCR. She wrote to the International Statistical Institute, asking if they had any old publication copies that they could spare. She offered to pay shipping and handling charges.[30]

David was demonstrating success at UCR. In 1972, her dean wanted to increase her salary, based on his recognition of her "meritorious and distinguished service." He wrote to Neyman, saying he had the "pleasure of recommending" David and asking Neyman for a letter of evaluation of David's scholarship.[31]

Neyman responded with this evaluation:

> It is a great pleasure enthusiastically to support your recommendation that Professor David be advanced to the status of Professor Overscale in the Department of Statistics. I have known Professor David since about 1935 when she embarked on her doctoral studies under my guidance. As a first step in this direction there appeared a paper by "FND" [David] summarizing a certain subject on which at the time I lectured at the University College, London. This first step was soon followed by several others, this time with FND as the sole independent author. Over the years that elapsed, FND has been assiduously working and publishing on a number of interesting and important subjects, from bombing problems during World War II, to symmetric functions, to tests for homogeneity, to problems of ecology and epidemiology, to forest fires, etc. Last but not least, FND's hobby must be mentioned. This is the history of science, particularly of statistics and probability. She has several papers published summarizing the results of her digging in various archives from Paris to Rome.

---

[24] David, F.N.: Letter to Neyman, J. 11 January 1974.

[25] Flyer re Open House, Special Events, 31 May 1974.

[26] David, F.N.: Letter to Carver, H.C. n.d. Note: H.C. Carver was in Santa Barbara.

[27] David, F.N.: Letter to Stone, M. 24 January 1974.

[28] Neyman, J.: Letter to David, F.N. 16 January 1974.

[29] Itinerary for Neyman, J. and Scott, E.L. 31 May 1974–1 June 1974.

[30] David, F.N.: Letter to International Statistical Institute 28 January 1974. Note: We did not find a record of how the International Statistical Institute responded.

[31] Dugger, M.: Letter to David, F.N. 27 October 1972. Note: Mack Dugger was dean of the University of California—Riverside college of biological and agricultural sciences.

> In addition to the above purely scholarly activity, FND (unofficially but affectionately "Duchess") exhibited an enviable organizational talent. What I hear about her Department at Riverside is very impressive.[32]

Neyman liked his martinis.

## 9.2 Having Academic Stature

Shortly after the start of the 1973 new year, David reached out to someone in the UCL classics department. She wanted to help the daughter of a friend who aspired to study classics at UCL. She wasn't sure who to write to, because she admittedly had been away from UCL for "some years." When David explained the girl's strengths, she mentioned that she was, "for an American, very well read," which gives us a peek into David's opinion of Americans. David's letter also gives us a peek into David's opinion of UCR, when she told the UCL classics professor that she found the atmosphere at UCR to be "rejuvenating if somewhat hectic."[33]

The Berkeley Symposium on Mathematical Statistics led by Neyman was going strong. David had contributed a paper to the 600-page Vol. VI, *Effects of pollution on health*. She was given notice that the volume had just been published by the University of California Press [4].[34] David was happy with the series, but not with the bookkeeping of the press, and she spoke her mind: "Please do not send me any more notices such as the enclosed [about ordering copies]. I paid in advance for the whole series and very much resent your incompetent bookkeeping. This is the third time I have had to protest, each time about a different volume."[35]

David accepted an invitation to speak at Oregon State University in the statistics department seminar series. She indicated she would talk on the subject of "Diversity with applications to forests and lightning strikes."[36]

She contributed this abstract:

> Measurement of diversity. How may we describe a forest in terms of the kind of species and the abundance of each? How may we describe the pattern in time and space of a disease in a city? How may we compare the patterns of lightning strikes for two different years? How may we compare the densitometer readings obtained from aerial photographs of the same place at two different time points?
> A simple and somewhat obvious criterion is suggested, and a mathematical description of its properties, in so far as they have been worked out, is given. A new type of function, named the augmented monomial factorial symmetric function, is devised and suggestions for its manipulation are made.

This was a lot of ground to cover, and David knew she would need 3 weeks to do it, so she decided to focus on the most important practical problems. She offered "The field is, at present, wide open for anyone with any mathematical sense to join in."[37] When it came to fires, David made it clear that she studied fires in forests, not houses.[38]

---

32 Neyman, J.: Letter to Dugger, M. 31 October 1972.

33 David, F.N.: Letter to Handley, E.W. 8 January 1973. Note: E.W. Handley was in the University College London classics department.

34 LeCam, L., Neyman, J., and Scott, E.L.: Letter to David, F.N. 15 January 1973.

35 David, F.N.: Letter to University of California Press 23 April 1973.

36 Calvin L.D.: Letter to David, F.N. 23 February 1973. Note: Lyle D. Calvin was chair of the Oregon State University department of statistics.

37 David, F.N.: Letter to Calvin, L.D. 10 April 1973.

38 David, F.N.: Letter to Bowen Jr., J.V. 16 May 1973. Note: J.V. Bowen, Jr. was in the University of Richmond department of mathematics.

David had her secretary phone Neyman and ask him to come to UCR to give a lecture.[39] He would receive an honorarium of 100 USD. David was in Berkeley for the weekend of 19–20 May, and she flew down to Riverside with him on Sunday evening.[40] On Monday, Neyman gave a talk to David's department on "Biased sampling and means of unbiasing. Example from astronomy."[41]

David continued to love books, both writing and reading them. In 1973, she accepted the position of Book Review Editor for *Biometrics*. Readers were invited to make suggestions to her about books to list, summarize, or review.[42] David wrote a lot of book reviews herself. She reviewed six books for one issue of *Biometrics* alone.[43]

The transition to David's book editor appointment was bumpy. Cambridge University Department of Medicine's A.W.F. Edwards thanked her for her "thoughtful and sympathetic review... Knowing your involvement with confidence theory I particularly appreciate your comments." However, David had not written the review. It was written by someone else under the previous editor. David told Edwards that she did in fact write a long complimentary review of his book, sent it in to the journal, learned of the other review, and withdrew her own review, but then the secretariat published the other person's review under David's initials. She apologized to the author of the review and then to the author of the book being reviewed.[44] David's professional honesty is noted.

Not all of David's book reviews were well received. An Australian author quarreled with her review of *Fundamentals of Biometry*. The author did not like that David used the Oxford English Dictionary's definition of biometry and wrote this to her: "While book reviewers are entitled to their opinions, I'm fortunate that your opinion expressed in the final sentence of your review is not shared by a number of teachers of biometry. The text is selling extremely well and many lecturers and professors have advised me that the text is most suitable for a class text."[45] Perhaps David wasn't hesitant to write negative reviews because she felt that no one read them anyway.[46]

There are many examples where David was a tough reviewer. She rated a National Science Foundation (NSF) proposal for weapon development as *Fair* when the only lower category was *Poor*, surely dooming the proposal. She argued that the proposer relied too much on the assumption of normality. More importantly, he had not convinced her of the usefulness of the project. "If he wants to 'sell' his methods he should at least come up with something new that has called for his proposed research work," she concluded.[47] For another example, she recommended that an article not be accepted for publication. In her judgment, the problems were not new and did not seem useful, and the authors did not know the literature. David chose not to recommend that the authors be invited to revise and resubmit their article.[48]

For yet another example, David was asked for her opinion of a paper that the RSS had solicited from an author. She "admired" the author, but found the paper to be "unnecessary." She wrote:

---

[39] Note for the file 30 April 1973.

[40] Itinerary for Neyman, J. 20 May 1973.

[41] Neyman, J.: Travel schedule, with notes, May–August 1973.

[42] Eisenhart, C.: Letter to David, F.N. 14 September 1973. Note: Churchill Eisenhart was at the US Department of Commerce National Bureau of Standards.

[43] Cady, F.B.: Letter to David, F.N. 25 August 1976. Note: This was for the March 1977 issue. Foster B. Cady was at Cornell University and was *Biometrics* editor.

[44] David, F.N.: Letter to Edwards, A.W.F. 10 October 1973. Note: A.W.F. Edwards was at the University of Cambridge department of medicine.

[45] Balaam, L.N.: Letter to David, F.N. 16 January 1974. Note: L.N. Balaam was University of Sydney, faculty of agriculture dean.

[46] David, F.N.: Letter to Dr. Harrington 19 January 1976.

[47] National Science Foundation: Proposal Rating Sheet 22 January 1976.

[48] David, F.N.: Manuscript review comments n.d.

1) As far as I can see [the paper] contains nothing new. This would not matter if the writer was able to give a new interpretation of known facts but this is not done. I cannot see anything which is not repetitive of previous work.
2) There is too much hind-sight. This is always a danger when looking back over scientific development. One tends to forget that at the time the person one is interested in was writing and working. He was at the boundary of knowledge with no very clear idea of how to proceed, and certainly no idea of what to do or he would have done it...

David didn't like the paper at all—thought it didn't present anything new, just a few corrections and a rehash—and was so bored by it that she told the person soliciting her opinion that he owed her at least two drinks for reading the paper! David thought that most of the paper should be "jettisoned." She asked "What is the RSS up to in asking for such a paper? One might describe the paper as a justification for having had a Guggenheim Fellowship and it should more appropriately be enshrined in a book than in a discussion before the Society. If the RSS has money to spare to spend in this way maybe it would be better to reduce the subscriptions and cut out this sort of thing."[49]

Samuel Kotz published a book in 1974, *Probability Theory: A Historical Sketch*, with L.E. Majstrov. David asked Kotz for a copy so that she could write a review in *Biometrics*. She wrote "I wish wholeheartedly that more scholars of religious literature would write monographs about what they know. Only then, I feel, will we start to get an insight into what, to me, is the ultimate puzzle—the duality of purpose in using dice (or astragali) for street gambling and for religious ceremonies. But such scholars should throw hindsight out of the window."[50]

The books that David wrote continued to attract attention and from unexpected places. For example, someone who was in the business of trouble shooting on refrigeration and air conditioning equipment for a grocery store chain wrote to her and said that he had used one of the methods in her *Games, Gods and Gambling* book and that it had worked well. The same person liked to play at the crap tables at Lake Tahoe and told her he had used some of the methods in his gambling hobby. He even asked to purchase copies of her lectures on gambling.[51] For a more mainstream example, the authors of a new textbook, *Statlab*, asked her permission to use material from her book. The requester told David he had read her book "with a great deal of pleasure"[52] and thanked her for it. David had been on a summer project with this person, and thinking back about it gave her "a good laugh." She gave the author "carte blanche" to quote what he liked.[53]

In November 1973, David was on her way to London for some personal business that she considered to be urgent. She expected to return to Riverside before the end of the year. On her way out the door, she sent a letter to a UCR colleague who was asking for a recommendation about a merit increase. David didn't have time to write a long explanation, but she was not shy to speak her mind: "I would be sorry if he went but one cannot (or at least I cannot) be held to ransom on what is an academic matter."[54] David's honesty and integrity are again demonstrated by this response.

Also on her way out the door, David sent a letter to a Berkeley colleague who asked her opinion about an article he was considering for a journal. She told him that she would support his judgment, but she would have preferred to see "some definitive theoretical approach rather than a lot of

[49] David, F.N.: Handwritten letter to Dr. Freeman 22 November 1977.

[50] David, F.N.: Letter to Kotz, S. 22 October 1974. Note: Samuel Kotz was at the Temple University department of mathematics.

[51] Hemphill, W.F.: Letter to David, F.N. 1 October 1973.

[52] David, F.N.: Letter to Krech, D. 14 January 1974.

[53] David, F.N.: Handwritten letter to Krech, D. 22 January 1974.

[54] David, F.N.: Letter to Kogiku, K.C. 6 November 1973. Note: K.C. Kogiku was UCR Department of Economics chairman.

computing... I hope this isn't the first swallow of a flight of empirical computing papers."[55] David's perspective was that data should be analyzed within a model (theoretical framework).

David may have been passed over at UCL, but she continued to be held in high esteem there (with the probable exception of Lindley), and she continued to have friends there. In December 1973, she received an invitation to return to the UCL department of statistics (which was now the department of statistics and computer science and would soon be headed by Mervyn Stone) as a distinguished visitor in 1974–1975. She would have a teaching assignment, which they described in some detail in the invitation. Her teaching would include an advanced MSc course of her choosing, a few case studies to third year undergraduate statistics students, and a series of lectures on the History of probability and statistics for the general education of staff and students.

UCL wanted to have David as a visiting professor for an entire term, which would have logistically been possible for her given that she still had the 28 Torrington Square apartment and the Graffham cottage. In addition, this would have suited her personal life as she would have been able to live with Joyce Townend instead of carrying on a transatlantic companionship. Ironically, what made this offer possible was that Lindley would be on a leave of absence, which freed up funds for the department to hire distinguished visitors like David.

The letter of invitation came from Mervyn Stone, who had studied under Lindley at a time when Lindley was a "eventually renowned teacher" at Cambridge. Stone was apparently not a Bayesian evangelist. His obit says this about how he got to UCL: "Dennis Lindley had recently been appointed to the Chair in Statistics at UCL and had a vision of promoting the Bayesian philosophy. Dennis was delighted to appoint Mervyn, who, he told us, was a Bayesian. But Mervyn was far too free-thinking to be limited by that epithet and in fact very little of Mervyn's work in statistical theory was directed toward developing formal Bayesian methods." In fact, Stone even published criticism of formal Bayesian methods [5].

UCL's offer was enticing to David: "Your connections with the Department only serve to emphasize how pleasant it would be for you to be with us for a period." Stone's letter of invitation also included the names of statisticians who would be present in the environment: "I was not familiar with the London Statistics scene when you were at University College but there has been, over the last few years, an impressive increase in attendance and inter-Collegiate Seminars and postgraduate lectures in Statistics, while the number of floating visitors seem to be increasing likewise. I trust, therefore, that you would find it interesting to be here. Likewise, there are many in London who would be delighted to renew their old acquaintance with you."[56]

David was, of course, interested in visiting the department of statistics and computer science at UCL.[57] She was still commuting between Los Angeles and London. She sent Stone a bibliography and other materials. She told him she was used to a 6-hour-a-week teaching load, and she could teach "probably, anything" that he would like her to, but she specifically mentioned Combinatory theory and History of probability. She checked at UCR and learned that there would likely be no objection for her to have a leave of absence. She had some concerns about finances, which she expressed to Stone saying that she would like some time to think over his offer.[58] But the invitation was kismet.

At the close of year 1973, David was invited to give a 45-minute lecture in a symposium to honor Neyman. It was organized by the Mathematical Institute of the Polish Academy of Sciences and

---

[55] David, F.N.: Letter to Tarter, M. 7 November 1973. Note: Michael Tarter was a University of California—Berkeley school of public health professor.

[56] Stone, M: Letter to David, F.N. 3 December 1973. Note: Mervyn Stone was University College London department of statistics and computer science professor.

[57] Stone, M.: Letter to David, F.N. 14 December 1973.

[58] David, F.N.: Letter to Stone, M. 24 January 1974.

the International Association for Statistics in the Physical Sciences. They would pay for her hotel and expenses in Warsaw, but she would need to pay the cost of getting there. David was one of 33 renowned statisticians from across the globe who were invited to participate.[59] But she had to decline the invitation. The symposium was too close to the beginning of the spring quarter at UCR when she needed to be on campus.[60]

March was a sad month for David. She lost her dear friend Joyce, who was her companion for 40 years. David told Blackwell, who in 1974 was in residence at the Study Center of the University of California, United Kingdom, and Ireland:

> I have been a great many times through London since we last met but not in circumstances when I felt like talking to anyone. Joyce, with whom I shared a home base for forty years, developed a cancer in November last year and died in March. I went backwards and forwards from L.A. to London more times than I like to remember. It was (is) a sad time, particularly since my retirement plans were based on the premise that we would be sharing a house in the South of France. Now I'm not sure what I want to do.

David sent Blackwell her love, after she commented that London seemed "to get seedier every time I pass through."[61]

David was equally prepared to offer personal and professional opinions. For example, she told Neyman how to handle disputes about the accuracy of the National Academy of Sciences (NAS) reports:

> We have discussed NAS expulsion before, and it is one of the few matters we disagree on. My own feeling is that the main remedy for lying is to provide opportunities for telling the truth. For instance, if one NAS member thinks an NAS report is inaccurate (deliberately or not), maybe NAS should be willing to print a five-page statement from him and mail it to all those to whom it mailed the report. The dissenter could of course include in his statement directions on how to get a more complete statement.

NAS had "already gone much too far in the 'truth-certifying' business" for her taste.[62]

David's opinion on affirmative action was valuable and influential on a national level in the United States. The Office of the Secretary of the Department of Health, Education, and Welfare in Washington DC appreciated receiving a copy of the comments on affirmative action that she sent to the University of California Vice-Chancellor Michael Heyman. They told her they found her comments very useful in their review of the University of California's affirmative action program.[63]

David summed up her affirmative action activities in a letter to the president of the California Women in Higher Education, Irvine Chapter. The letter is worth repeating here in its entirety:

> 1) My own campus (Riverside) has a cooperative and willing administration. I am chairman of the Chancellor's Committee for Affirmative Action (Academic), chairman of the Chancellor's Committee on the Status of Women (Academic and Staff) and a member of the Academic Senate Committee for Affirmative Action. Any data which I need on salaries etc. is provided by the Administration who seem to be several steps ahead in their thinking when compared with the academic faculty. In my experience any block in the advancement of women at any level (graduate student to faculty) comes from the moribund attitude of one's older male colleagues. This is true worldwide and not just in the U.S. I don't see how any Administration can deal with this except by strict adherence to the Affirmative Action Guide Lines, which is usually done.
> 2) The Academic Vice-President (Angus Taylor) appointed a statistical committee to advise him and I was informed a few days ago that, in spite of the reorganization now taking place at University Hall, the statistical

[59] Bartoszynski, R. and Olech, C.: Letter to David, F.N. 19 December 1973. Note: Robert Bartoszynski was organizing committee head and Czeslaw Olech was director of the institute. The list of persons who had been invited to participate was attached to the letter.

[60] David, F.N.: Letter to Bartoszynski, R. 30 January 1974.

[61] David, F.N.: Handwritten letter to Blackwell, D. 4 December 1974.

[62] David, F.N.: Handwritten letter to Jerry 9 May 1974. Note: Jerry was presumably Jerzy Neyman.

[63] Holmes, P.E.: Letter to David, F.N. 13 December 1974. Note: Peter E. Holmes was Department of Health, Education, and Welfare, Office of Civil Rights director.

> committee would continue in being. The committee consists of Elizabeth Scott (UCB), Chin Long Chiang (UCB) – both of whom are my close friends – and myself. We had a busy time all through the last Christmas recess advising the University Administration regarding the UCB submission to the criticisms of HEW. As you may know UCB accepted the concept of parity and the salaries position is (I think) still under discussion. [Parity means percentage of women faculty in a particular discipline = percentage of women PhD's in that discipline.] If you have read the recent furore raised by a member of the Mathematics Department at UCB – it reached most newspapers – you will recognize it for a little of the faculty opposition of which I wrote earlier.
>
> 3) That the University-wide Administration is not unmindful of the salaries position is shown by the fact that it has already been investigated for UCLA by Dr. Saxon when he was Vice-Chancellor there. A similar investigation was made by our Statistical Committee last Christmas for U.C.B. and yet another investigation for U.C.R. is going through the computer. [Have you done one for U.C.I.? If so let me know.] There appears to be a desire on the part of all our U.C. administrations to become appraised of the true position. I do think however that action will have to depend on faculty attitudes.
>
> 4) The person at University Hall who has the charge of Affirmative Action is Gloria Copeland who is the first woman to reach the rank of Assistant Vice-President (Academic). If you have evidence which you want the Administration to know about I think she would be the right person to talk to.
>
> 5) I am sorry if I seem a bit negative in my attitude. As I said previously I think it boils down to the attitude of faculty members and this is something which probably cannot be overcome except by a process of attrition which involves a steady effort by women at maintaining a standard of excellence. The chief effort in the next five years should be directed at trying to get more women into graduate programs. Or so I think.[64]

As her letter indicates, David was heavily involved in the affirmative action efforts of the University of California. For context, we refer you to the book by Golbeck [6] which contains considerable detail about these gender and racial/ethnic equity efforts in academe at the time.

In August 1976, the University of California vice president for academic and staff personnel relations appointed a committee to review the university's practices for collection and retention of data, paying particular attention to affirmative action and ethnicity data. David was one of six people from across the University of California system to be appointed to this special committee. The committee was to review and make recommendations about the administrative authority for collection and retention of affirmative action and ethnicity data, review the purposes and use of such data, and develop policy to ensure confidentiality of such data. David was perfect for this "extremely important" assignment.[65]

Here is an example of a matter which was discussed in the committee:

> Kleingarten [sic] asked 2 children in his office to come up with a simple formula which was to be distributed to all campus regarding the number of years each department and staff personnel grade was going to take to achieve parity for women and minorities. The 2 children produced the appended but did not put numbers in it (then) to see if they make sense. The campuses started getting - sometimes - negative numbers and it was at this stage (Friday last) I was called in.
>
> The problem is ludicrous. It is necessary always to take the "best qualified person" for any opening so if a woman leaves the chances are she will be replaced by a man and so we go backwards. I wanted to get it somehow (1) that if a woman leaves she is replaced by another woman, (2) that if a man leaves he will be replaced by either a man or a woman. How else ever to get anywhere? I am meeting [UCB Biostatistics Professor Chin Long] Chiang at 3 this pm. Will keep you informed as to what we do. There is unusual urgency.[66]

David was known to become animated about issues such as these.

David saw affirmative action as a fight, and she found that the time and effort spent to fight were painful. She said "There's an awful prejudice in America, overt prejudice... academicians will always

---

[64] David, F.N.: Letter to Whiteley, R. 30 April 1975. Note: Rita Whiteley was California Women in Higher Education, Irvine Chapter, president.

[65] Kleingartner, A.: Letter to Pohl, I. 10 August 1976. Note: Archie Kleingartner was University of California vice president, and Ira Pohl was University of California—Santa Cruz professor.

[66] David, F.N.: Handwritten letter to unknown, 23 June 1976. Note: We don't know to whom this letter was addressed or what the attachment said.

go for [attack] a woman if they can, rather than a man. It's also true in England, but it's not overt. I had to deal with that prejudice. Female assistant professors would come to me and weep because I was on the President's University Committee for Affirmative Action." Later in life, David thought there was still prejudice, but she also thought there had been overall gains [3].

With all of this involvement in women's issues, David was clear about who were the "good" regents and who were not. When it came time to nominate regents, she weighed in, proposing that Regent Heller serves an additional term. "She has always impressed me as being 'with it' on the educational side and I have always found her approachable, well-informed and humane in her outlook," David wrote.[67]

The Association for Women in Mathematics was founded in 1971. It's mission, as currently stated, is "to create a community in which women and girls can thrive in their mathematical endeavors, and to promote equitable opportunity and treatment of women and others of marginalized genders and gender identities across the mathematical sciences." [7] David was a supporter of women, but she did not regard herself to be a mathematician. Nevertheless, she had become a member of the association by 1976.[68]

Berkeley continued to be a draw. David planned to spend August 1974 at the US Forest Service Southwest Range and Experiment Station outside of Berkeley.[69] In the next year, Lindley invited David back to UCL to help celebrate E Pearson's 80th birthday.[70]

Neyman is credited with the idea of the confidence interval. A confidence interval is an indicator of how good an estimate is, with smaller confidence intervals indicating more precise measurement. Confidence intervals are a tool in the frequentist toolbox, and the idea has been around for a very long time, since 1937.

In 1976, Bayesian Lindley launched an attack in the United Kingdom on the confidence interval. Frequentist David saved some of the discussion papers that were published in the RSS's *News and Notes* in reaction to Lindley's attack, including ones by E Pearson (who gave a warning that "it is better for none of us to make use of *News and Notes* for rather crude proposition of our particular beliefs") and others. A fervent frequentist, David saved these papers for so many years that they survived her passing.[71] Neyman commented to E Pearson on the attack, saying "We seem to be in a period of a flare-up of interest in the 'old-time theory'."[72] Incidentally, in the United States, one occasionally hears about "old-time religion,", but this was "old-time theory."

The dialog went on for months, and Lindley didn't want to let the matter go. He challenged E Pearson point by point, concluding with "What about a relevant reply to my original letter?"[73] E Pearson was offended by the whole matter and wrote to Neyman: "this collection may amuse you.

---

[67] David, F.N.: Letter to Maradudin, A.A. 26 January 1976. Note: A.A. Maradudin was in the University of California - Irvine physics department.

[68] David, F.N.: Handwritten letter to the Association for Women in Mathematics 4 January 1976. Note: The letter included a check for 5 USD to cover David's membership dues.

[69] David, F.N.: Letter to Ramachandran, G. 11 June 1974. Note: G. Ramachandran was at Borehamwood, Hertfordshire, Fire Research Station, Building Research Establishment, Joint Fire Research Organization, department of the evnironment and fire offices committee.

[70] Lindley, D.V.: Letter to David, F.N. June 1975. Note: D.V. Lindley was University College London statistics department head and professor.

[71] Letters by Lindley, D.V., Greenfield, A.A., Jackson, P.H., Pearson, E., Hamaker, H.C., Chatfield C., and Leonard, T. in *News and Notes* 1976.

[72] Neyman, J.: Letter to Pearson, E. 2 March 1976.

[73] Lindley, D.V.: Letter to Pearson, E. 12 May 1976.

I don't like biting the hand that feeds me, but Dennis Lindley seems to be going too far, and from various correspondence I have had the real working statisticians feel the same...."[74]

Next came a very blunt, lengthy, and biting letter from E Pearson to Lindley. E Pearson essentially came out of retirement to write it. He ended his letter by saying "I agree with Chatfield's advice that if you want to encourage people to take notice of you, you should make clear that what you say is 'a personal view based on your own philosophical approach to statistical problems... this, my reply, could be published, but that of course is impossible! Perhaps [this exchange] will find their way into one of our obituary notices as curios of the 1970's.'"[75] Maybe it didn't find its way into an obituary, but part of it did find its way into this book.

E Pearson kept Neyman informed about this dialog on confidence intervals. Neyman was appreciative and replied that he "was amused by the amount of nonsense that is now being published by respectable individuals."[76]

David served as a role model for future statisticians, in addition to her own colleagues and students. The Hugh O'Brian Youth Foundation featured David and the field of statistics. The executive director summed up David's remarks:

> In an interview, Professor David stated: "We have found that graduates who combine the ability to do applied statistical analyses with understanding of the vocabulary and technical requirements of a particular discipline go out to extremely well-paid jobs, positions of considerable power. For instance, we have had mathematics graduates who also specialize in hydrology, in agricultural economics, in plant genetics – in any of the biologies – go into positions where the companies, agencies or research institutes need to be able to predict the outcome of management techniques. One of our graduates is calculating the number of lemons to be expected on the trees of growers under varied conditions. Even historians, to work in the new field of cliometrics, need to understand applied statistics. I'd recommend that almost anyone would find him or herself better qualified for tomorrow's job market with a heck of a lot of statistics courses under the belt!"

This was part of the foundation's mission to highlight for youth different fields that offered outstanding job prospects. David was further quoted to say that there was "no shortage of jobs at any level and in any field of application in statistics."[77] The same is true today, almost 50 years later. The foundation article further explained:

> Statistics involves the theories and methodologies for the collection, presentation, analysis and interpretation of quantitative data. Statisticians work closely with other scientists and researchers to develop new statistical techniques, adapt existing techniques, design experiments and direct the analysis of surveys and retrospective studies. Some of the diverse fields in which statistical methodology has had extensive application include the disciplines of anthropology, archeology, history, library science, law and public policy.
> Professor David's belief in the high job availability is supported by an extract from the February 1976 issue of The Southern California Statistician: "About 25,000 statisticians are employed nationwide and about 1/3 of these are women. Approximately 2/3 work in private industry – manufacturing, transportation, public utilities, finance and insurance companies... another 20 percent work for the Federal Government... the final 15 percent have jobs with the state and local governments or in colleges and universities. At a national level, a minimum of 290,000 and probably well in excess of 1/3 of a million individuals now work and/or are paid to do work in this country which involves the development, manipulation, and/or application of statistical methods, procedures, and data. The outlook at a national level shows that approximately 1,700 statisticians should and probably will be hired each year for the rest of the decade. Primary job sources will be in the fields of trend forecasting and modernization of accounting procedures... The overall outlook is that the job market will be generally good if the training and knowledge of statistics is combined with another field of application..."[78]

---

[74] Pearson, E.: Handwritten letter to Neyman, J. 13 May 1976.

[75] Pearson, E.: Letter to Lindley, D. 17 May 1976.

[76] Neyman, J.: Letter to Pearson, E. 19 May 1976.

[77] Gorham, S.: Letter to David, F.N. 5 October 1976. Note: Sophia Gorham was Hugh O'Brian Youth Foundation (in Los Angeles) executive director.

[78] Career Emphasis, copy for Ambassador, 1, 1, n.d.

David described the field of statistics as one that was "rapidly growing."[79] According to the US Department of Labor, statistics today is still one of the fastest-growing occupations in the United States.

## 9.3 Retiring from UCR

All the while at UCR, David kept a foot in the door at UCB. For example, Neyman engaged David to teach at UCB in 1972. Her assignment would be a five-credit-hour introductory statistics course in the first summer session. She would lecture 5 days a week, and there would be six laboratory sections attached to the course that would meet 5 days a week.[80] Scott was departrment chair of statistics at UCB at the time. She informed David that her summer teaching salary would be 4,369 USD, paid in two installments.[81]

David described her History of probability class, which she taught both at UCR and UCB, as follows:

> The course is entitled History of Probability and I am concerned with trying to explore how probability (chance, luck) developed and to try to trace its origin back to religious rites and early game playing. Since in its earliest origins there (probably) was no one who could count or read or write I begin with archeological evidence and continue through the references in the classical literature and subsequent mediaeval writings until I get to solid ground with enumerations and calculations. I doubt if the course could ever be described as mathematical. I suppose it could be said to be of general cultural interest to most disciplines.
> In order to ensure that students do some work I ask for a project to be carried out and written up, the subject of the project having some bearing on some facet of my lectures.

David allowed students to work in groups and turn in a group project. She allowed them to take the course as Pass/Fail.[82] The course was in the statistics catalog, but at least one student argued that it should be counted as satisfying a university history requirement.[83] David also tried to help students get the course accepted as a breadth requirement.[84]

David's course caught the attention of those outside the world of academia, including the gambling community in Las Vegas. The local newspaper covered her work on gambling, and the local library invited her to submit any reports and materials that could be included in their new special collection on gambling.[85] David's expertise was also sought after by scholars outside of the field of statistics. For example, a French literature professor at the University of California—Los Angeles, having seen one of David's book reviews, wrote to her asking for specific sources from the orthodox classical literature, Greek and Latin.[86]

Former students raved about the high quality of David's teaching and advising, even though David at this point loathed teaching. Also at this point, she disliked teaching in retrospect, saying that "In

[79] David, F.N.: Letter to Kruskal, W.H. 13 January 1977. Note: William H. Kruskal was in the University of Chicago department of statistics.

[80] Department of Statistics Statistical Laboratory SUMMER SESSIONS SCHEDULE 2 February 1972.

[81] Scott, E.L.: Letter to David, F.N. 28 February 1972. Note: Elizabeth L. Scott was University of California–Berkeley statistics department chair.

[82] David, F.N.: Letter To Whom It May Concern 22 April 1974.

[83] Hailand, J.B.: Letter to David, F.N. n.d. Note: The student was Jeffress B. Hailand.

[84] David, F.N.: Handwritten letter to University of California–Berkeley dean 30 April 1976.

[85] Anderi, S.: Letter to David, F.N. 21 June 1974. Note: Susan Anderi was University of Las Vegas special collections department librarian.

[86] Harrington, T.M.: Handwritten letter to David, F.N. 13 January 1976. Note: Thomas M. Harrington was University of California—Los Angeles professor of French literature.

my young days it was a necessary irritant to me to get on with my research work." [3] For example, one former student wrote "While I know I shall remain one of the thousands of nameless faces (or faceless names) with whom you have shared your fine expertise, I shall never forget your fantastic Bio-Statistics course. The information continues to be a valuable tool for me."[87] For another example, a former student wrote: "Hope you are well and raising hell as usual. [para] Just wanted to tell you again how much I appreciate the help you were to me during the last few months. This job seems kind of tough at the moment, but I think this is the right direction to be heading in. Probably would be in a state of semi-retirement as a political science prof—somewhere now without the moral encouragement you gave me. Looking at your Department from the inside for a few months gave me some feeling that this kind of research is a natural state of affairs and not whatever it is that social scientists usually fear. So I feel better equiped [sic] to do the work here. Thank you again—and I'll try to pass it along as you always tell me to do."[88]

Students from other institutions knew of David and requested her assistance in learning statistics. For example, an instructor in sociology at a community college was one of David's former students. The instructor knew from personal experience that David had written a great study guide for elementary statistics. One of the instructor's students was getting ready to take a statistics course. The student asked David for a copy of the study guide and was willing to pay for it.[89] David didn't ignore the student's request, but she couldn't fulfill it as requested. The materials were actual hand-outs and there were too many. They amounted to "considerably more than a study guide." So David asked the student for a list of topics, so that she could consider sending the hand-outs that corresponded to those topics.[90]

In 1974, the UCR department of statistics as led by David continued to have a strong reputation. In fact, UCR was competing with UCB for graduate students.[91] David thought that students applying to both places—if financial incentives were the same—would want to go to Berkeley "where the action is."[92] However, there was enough money at UCR to invite visiting professors to teach for several quarters at a time.[93]

David was in London for a short time at the beginning of 1977.[94] She wanted to go back to London in the summer and was negotiating temporary working space at the University of Oxford's Mathematical Institute. She was looking for libraries in the United Kingdom that were broad enough in scope that they would have what she needed. She wanted to:

> …trace the impact of the social unrest in France in the 18th century on mathematical thought. At the beginning of the century we have Daniel Bernoulli and jurisprudence, and de Moivre and gambling. As the century wore on the shift was gradually but unmistakably away from probabilities of gambling and more and more toward those of social problems, and I think the development of probability theory was to a certain extent dictated by the desire to get answers to them. I hope to find a good deal of the material I need in the Oxford libraries.
> Another pet project I have will probably amuse you. I have long been impatient with emphasis in modern statistics which is placed on the normal distribution. I have in mind to write a text in which I make a minimum

[87] Belling, B.: Handwritten letter to David, F.N. 11 August 1975. Note: Bronwyn Belling was at the San Francisco Development Fund.

[88] Pam (last name unknown): Handwritten letter to David, F.N. n.d.

[89] Alesso, N.: Handwritten letter to David, F.N. 22 April 1974. Note: Nicholas Alesso was a US customs inspector at the Los Angeles International Airport.

[90] David, F.N.: Letter to Alesso, N. 25 April 1974.

[91] Blackwell, D.: Handwritten letter to David, F.N. 25 November 1974. Note: David Blackwell was a University of California—Berkeley, department of statistics professor.

[92] David, F.N.: Letter to Blackwell, D. 5 December 1974.

[93] David, F.N.: Handwritten letter to Neil (last name unknown) 12 February 1976.

[94] David, F.N.: Letter to Archbold, J. 3 January 1977. Note: J. Archbold was *Mathematika* secretary.

> of assumptions in order to interpret the data of experience and, if possible, leave out the shadowy background of the normal population altogether. It is possible to go a very long way on just the assumption of randomness.

David planned to leave Riverside in the summer and use her house in Graffham as her headquarters for this work.[95] As it turned out, David decided she would leave UCR "for good" at the end of June and never did get around to writing the book.[96] Her "G.G.G." book —as UCB Statistics Professor Lucien LeCam called her *Games, Gods and Gambling* book—continued to draw interest, and people's copies were known to be borrowed only never to return.[97]

As the academic year 1977 began, David knew it would be her last year at UCR. She had been there for 10 years, since 1967, and she was 68 years old. She would retire.

But David continued to be intellectually active. She was interested in many things, among them the space-time interaction of diseases. She explained to a colleague at the California State Department of Public Health:

> I have been interested for some years in the space-time interaction of various diseases. The cancer data has been difficult because of the reporting by counties. I think I have now solved this difficulty just in time to hear you say that there is California data available by census tract. I want to look at the association between type of cancer and (i) other types of non-cancer deaths (ii) time (iii) some sort of measure of social and economic status (iv) average yearly air pollution (v) some sort of measure of water mineral content. There may be other factors that you might suggest. The point is that given the different types of death have been recorded on tape or cards or some such, it will be easy to add other factors and just do a computer run. The main thing is to get hold of the data.

She would need a student research assistant.[98]

David found papers on these topics and said they led her to be interested in lattices and maps. She sent the papers to her colleague, along with a three-page explanation of the mathematics involved and where she would be headed in her research on the matter. She believed her work would be useful in the field of epidemiology.

David had done much work for the Forest Service. She started to work for them as a consultant in 1963, and by 1977 she was "... spending a third of each year on forestry studies chiefly connected with the spread of forest fires (not unrelated to her civil defense studies during the war and in some cases directly related to nuclear war scenarios)." [8]

The work proposed now was different because it involved humans instead of trees. However, David was eager to accept the challenge. She spelled everything out for her colleague, with an apology: "Sorry to spell all this out. It will have been clear to you that my difficulty in the past, as far as humans are concerned, is that the data are reported for irregular spaces. For a forest I don't have this difficulty since the data can be obtained for 1/4 acre plots." She then concluded: "I have now overcome my irregular space trouble. Thus I can consider an arbitrary array of blocks [she included an example drawing] and work out a diversity score. So I think I can look at the interpretation of clustering on a map and on a time succession of maps."[99]

David also was interested in working on sickle cell anemia as a statistical consultant to The Aerospace Corporation. She said she would be happy to join the group, because in recent years

---

[95] David, F.N.: Letter to Kingman, J.C. 9 February 1977. Note: J.C. Kingman was a University of Oxford mathematical institute professor.

[96] David, F.N.: Letter to Mann, N.R. 17 February 1977. Note: Nancy R. Mann was at the Rockwell International Science Center.

[97] LeCam, L.: Letter to Irwin, J.M. 23 February 1977. Note: Lucien LeCam was a University of California Berkeley statistics professor, and Jack M. Irwin was in Albuquerque, New Mexico.

[98] David, F.N.: Letter to Goldsmith, J.R. 23 February 1977. Note: John R. Goldsmith was at the California State department of public health.

[99] White paper on the space-time interaction for diseases, three pages 24 February 1977.

most of her work had been "concerned with the interpretation of biostatistical modelling."[100] And she was engaged in writing a monograph on randomization techniques, which she expressed interest in publishing with Prentice-Hall, Inc.[101]

Both David and E Pearson continued to be interested in researching and writing about history after their retirement, and this concerned E Pearson. He wrote to her: "As I have heard from various sources that you are planning to get down to historical research and writing, I think I should let you know where I stand on the matter, to avoid any chance of treading on each other's toes."[102]

Once David's retirement plans were made known, she received letters of inquiry for her position.[103] A search committee was formed. It was chaired by R.T. Wedding, a professor in the UCR department of biochemistry. There was a lot of smog in Riverside, California, with no prospects of appreciable improvement in the near future. At least one potential candidate removed himself from consideration because of the air pollution. The same person commented: "You have a big job in filling Professor David's shoes."[104]

David did retire from UCR at the end of the 1977 academic year. Her department granted her the status of emeritus professor, which was an honor that came with the expectation that she would continue to contribute to the profession.

There was a retirement celebration at UC—Los Angeles on 16 May 1977. Mathematician/statistician Thomas S. Ferguson from UCLA gave introductory remarks. Neyman gave a speech, Highlights of Professor David's Career.[105] D.V. Gokhale gave a talk, Professor David's Contributions at Riverside. Then David gave some remarks. Finally, there was a session, Recent Contributions to Research Areas Pioneered by Professor David. Chaired by David, this session had three papers: (1) G. Haggstrom, Rand Corporation, Optimal Designs in Covariance Analysis; (2) J. Rice, UC—San Diego, An Application of Shot Noise; and (3) M. Sobel, UC—Santa Barbara, Dirichlet Distributions and Waiting Times. Then there were cocktails and dinner at the Branding Room Restaurant in Santa Monica.[106]

David turned around and joined the UCB Graduate Group in Biostatistics, which was being chaired by her friend Chin Long Chiang. There she intended to work on the space-time interaction of diseases that included cancers with similar treatments as sickle cell anemia.[107] She also continued her social life in Berkeley, where she frequently went to the Sunday midday dinners prepared by Scott at Neyman's house [9] (p. 132). University of Wisconsin Statistics Professor Emeritus Yandell recalled having a meal at a restaurant with Neyman, Scott, and David when he was a graduate student at UCB: "Neyman was in the middle and could not get a word in edgewise as Scott and David went on and on. An unusual situation for him!... [David] was always friendly and supportive of me and other students"[108]

---

[100] David, F.N.: Letter to Maxwell, F. 10 March 1977. Note: Floyd Maxwell was at The Aerospace Corporation in Los Angeles.

[101] David, F.N.: Letter to Feyen, P.J. n.d. Note: Paul J. Feyen was Prentice-Hall, Inc. biology editor.

[102] Pearson, E.: Handwritten letter to David, F.N. 11 December 1977.

[103] Rolph, J.E.: Letter to David, F.N. 16 September 1976. Note: John E. Rolph was at the Rand Corporation.

[104] Rolph, J.E.: Letter to Wedding, R.T. 16 September 1976. Note: R.T. Wedding was a University of California—Riverside biochemistry professor.

[105] Neyman, J.: Travel 1977.

[106] Notice: Special Colloquium to honor Professor F.N. David, Chairperson, Department of Statistics, University of California—Riverside, Monday, May 16, 1977, University of California, Los Angeles.

[107] David, F.N.: Letter to Maxwell, F. 10 March 1977.

[108] Yandell, B.: email to Golbeck, A.L. 10 May 2017.

Up to a few months before her departure from UCR, David was still providing professional advice and recruiting people to her faculty. She warned a colleague to not reveal too much in a grant proposal, because "it might give other people ideas." She told the same colleague she had applied for them to have a promotion to associate professor with tenure at UCR: "I hope I have succeeded in my efforts. I think you are worth more as a colleague than the sum of several others."[109]

David was intellectually generous as a chair. She was willing to help her junior faculty members by discussing statistical matters with them, even when she was not directly involved in the research.[110] She remembered birthdays.[111]

David was succeeded as chair by S. James Press, a Bayesian. It was as if the Bayesians were on David's tail. Things reportedly changed "very rapidly" at UCR after that.

In fact, one could say that it became a dark time at UCR.

One of David's hires received a "strongly negative" performance appraisal. They were told by the new chair that the review would be favorable, but it didn't turn out that way. According to another faculty member, it was the new chair who made the case for the negative review. This was a blow to the faculty member with the negative review, because they had always assumed "the people I was dealing with were fair and open. If they aren't, I'd rather not have anything to do with them. It is an unpleasant thought that you might have to defend everything you do... It is a real pain to have to worry about this mess that is unrelated to anything useful or interesting." The faculty member told David "Pretty soon all the nice people will be gone... It certainly looks like Riverside will be easy to leave."

Another faculty member thought the new administration "wanted to remove all non-tenured faculty and others to fill the slots with his own people...."[112] A young assistant was fired. The department was given three new positions in addition to two that were about to be vacated, which seemed inexplicable, because it was at a time when the campus supposedly was "going down."[113]

The faculty member who received the negative performance appraisal moved to another university where he had a long and successful career. A year after he left UCR, he wrote to David: "I am now more comfortable in the Entomology Department than the Statistics Department at UCR. Most of the people in the Stat. Dept. are friendly toward me, but they certainly weren't willing to risk anything to help me. Timid friends may be pleasant, but they are not very reliable."[114] In the same letter, and knowing David's strong negative feelings about Bayesian statistics, he reported the following amusing story: "I talked about these 'Bayesian birds' in Mr. Neyman's seminar in April 1979. At Mr. Neyman's the evening before David Blackwell offered a toast to 'Bayesians, feathered and otherwise.' I like to comment that birds should be Bayesians, they have the brains for it." He followed the story with three pages of statistical ideas, in which he mentioned that "Bayesian methods are more appropriate for foraging than for inference" and that he should be so busy as she was in "retirement."[115]

Another person elected to leave the department of statistics and transferred to economics. One of David's closest faculty members wrote: "If Press wants to remove all non-tenured faculty and others

---

[109] David, F.N.: Letter to Beaver, R.J. 3 May 1977. Note: R.J. Beaver was in Albuquerque, New Mexico.

[110] David, F.N.: Letter to Buerger, A.A. 17 May 1977. Note: Alfred A. Buerger was at the University of California—Irvine California college of medicine department of physical medicine and rehabilitation.

[111] Sevacherian, V.: Handwritten letter to David, F.N. 23 November 1977. Note: Vahram Sevacherian was at the University of California—Riverside.

[112] Sevacherian, V.: Handwritten letter to David, F.N. 23 November 1977.

[113] Green, R.: Handwritten letter to David, F.N. 13 December 1977. Note: Richard Green was at University of California—Riverside in the statistics department.

[114] Green, R.: Handwritten letter to David, F.N. 13 May 1981.

[115] Green, R.: Handwritten letter to David, F.N. 13 May 1981.

to fill the slots with his own people, he is probably going to get it all sooner than he realizes. [para] I think I better stop writing about UCR because I notice myself getting progressively more and more angry as I continue."[116]

In 1971, UCR gave David an Academic Senate Distinguished Teaching Award. This award "recognizes exceptional effort and achievement in teaching by a professional faculty member." [10] The award is perhaps more meaningful when one realizes that the students she was teaching at UCR were adult learners, including many Vietnam War veterans. After David retired from UCR, she taught at UCB for 10 more years and continued her consulting work for the US Forest Service [1].

## References


1. Garber, M.J., Gokhale, D.V., Utts, J.M., and Beaver, R.J.: In memoriam: Florence Nightingale David, statistics: Riverside (2022) http://texts.cdlib.org/view?docId=hb238nb0fs;NAAN=13030&doc.view=frames&chunk.id=div00019&toc.depth=1&toc.id=&brand=oac4 Accessed 4 October 2022.
2. Daly, J. (updated by Smidt, B.): A brief history of the department and the statistics major, Statistics Department, Cal Poiy (2022) https://statistics.calpoly.edu/content/brief-history-department-and-statistics-major Accessed 27 July 2022
3. Laird, N.M.: A conversation with F.N. David, Statistical Science 4(3), 235–246 (1989)
4. David, F.N.: Measurement of diversity: multiple cell counts. In: LeCam, L.M., Neyman, J., and Scott, E.L. (eds.) Proceedings of the Sixth Berkeley Symposium on Mathematical Statistics and Probability, Volume I Theory of Statistics - Multivariate Analysis - Population Studies and Branching Processes, University of California Press, pp. 109–136 (1972)
5. Galbraith, R.: Memories of Mervyn Stone, (2021) https://www.ucl.ac.uk/statistics/sites/statistics/files/meryvn-stone-obituary.pdf Accessed 29 July 2021
6. Golbeck, A.L.: Equivalence: Elizabeth L. Scott at Berkeley. Boca Raton, FL: Chapman and Hall/CRC Press (2017)
7. Association of Women in Mathematics (2022) https://awm-math.org/about/ Accessed 15 August 2022
8. Barton, D.E.: Obituary: Florence Nightingale David, Journal of the Royal Statistical Society A 157(2), 299–303 (1994)
9. Reid, C.: Neyman-From Life. Springer-Verlag (1982)
10. University of California - Riverside: Academic Senate Awards: Distinguished teaching (2022) https://senate.ucr.edu/awards/ Accessed 4 October 2022


[116] Sevacherian, V.: Handwritten letter to David, F.N. 23 November 1977.

# Part VI

# Down the Memory Lane

# Remembering 10

*. . . David. . . was. . . a general mathematical bad ass who regularly took on the patriarchies' "this is a man's world" nonsense.*

Sara [last name not given]: Deeply Trivial web site (2017) [1]

## 10.1 Colin Mallows Interview (10 September 2018)

I spent most of my career at Bell Labs.

In my second year as an undergraduate in the mathematics department at UCL, I was required to take a few lectures outside my department. It was purely by chance that I picked statistics, and I happily sat through lectures from F.N. David. . . David was a very inspiring lecturer. I was surprised to learn much later from an interview published in *Statistical Science* that she didn't enjoy teaching [2], but she certainly conveyed the attraction of the discipline. . . That was my first contact with statistics, and David turned me on.

FND used to lecture on probability. She used her own book. . . UCL had a very stellar faculty, and David was part of it [3]. It was David's lectures that inspired me to transfer from the math department to the statistics department for my final year as an undergraduate.

My work as a PhD degree student at UCL in 1951–1953 was supervised jointly by David and N.L. Johnson, both of whom later emigrated to the USA. They gave me a sequence of four problems to work on, each of which resulted in a part of my 1953 thesis, and each of which resulted (eventually) in a published paper. . . . In my thesis, I thanked "my supervisors, especially Dr. F.N. David, for their continual advice and encouragement." My recollection is that I met with David and Johnson every week. . . . From 1953 to 1955, I was conscripted into the British Army. So after I left UCL in 1953 for the Army, there was no contact for 2 years, and when I returned to the staff in 1955, I was still very junior and did not have much contact with her.

I knew David was interested in her relations with Berkeley and that Johnson was getting interested in leaving England (he told me once that he saw no future there), but I never talked to her about that. Many years later I wrote to her to apologize for not having thanked her properly for all she did for me in supervising my thesis, and I got back a nice letter.

I admire David's continual output of research papers. I think this "rubbed off" on me since I have been trying all my life to complete the next paper and get it published. . . . Of course, her output of



A. L. Golbeck, *Florence Nightingale David*,
https://doi.org/10.1007/978-3-031-77768-4_10

published papers and her many books are important. But I would say that the most important thing was her influence on her students.

## 10.2 James Baldwin Interview (8 May 2019)

I received a bachelor's degree from UCR, and then I went on to complete a PhD degree at UCR in statistics. I spent most of my career at the U.S. Forest Service. I am retired, but I liked the forest service work so much that I am a volunteer at the research station where I worked.

I first met David when I was a graduate student in statistics, and David was chair of the department. She had been at UCR for 2 or 3 years. I knew of her when I was an undergraduate, but I never actually met her until I became a graduate student in her department. David had a reputation among the undergraduates. She started every course by saying "You might notice that I speak differently from most of you. I speak the pure King's English."

I was a teaching assistant under David. Being a teaching assistant for her was really good for me. Her teaching assistants would grade the papers. Students would occasionally come to us and say we graded incorrectly, and they were going to go talk to David about how poor a job we did. We'd say "Great! Don't say we didn't warn you!" David was very supportive of her teaching assistants, and she'd just chew up the complainers. People would come back and say "Why didn't you warn us?" And we'd say "But we did." David was fierce.

I was David's head teaching assistant for about two and a half years. I created the initial draft of the exams. Some of the questions David added were more general like the following: Here's some data, what do you think the correlation coefficient might be, and is this what you got? People would write things like "I thought the correlation coefficient would be around 5, and I calculated the correlation coefficient to be 2.5, which isn't close to what I thought it would be." Now, of course, correlation coefficients range between − 1 and 1 (they certainly can't be larger than 1), and so some people clearly were clueless about the range of the correlation coefficient. Not everyone in the class listened, so some of the student answers were hilarious. After the exams the other teaching assistants and I did all the grading (David didn't do any of the grading). In fact, we teaching assistants got together with a bottle or two of wine while grading. The wine made the student answers seem even more funny.

When I was at UCR, David taught no advanced seminars or other advanced courses. She just taught the introductory course, basic statistics 101. She was really tough, but she gave good lectures, and her classes were well attended (her classes were always packed). She had a very loud voice that was very clear (there was no mumbling at all). She was very good at explaining the concepts and told stories throughout that kept people interested. I'm not aware of any overt problems in the classroom about her being a woman. Any male who would think he could challenge her would just get chewed up. Any challenges were stopped very quickly. Had she been less intimidating, she might have had issues, but she was intimidating enough at the right times to stop things from getting out of hand.

The first time I really interacted with David was when she was the chair of the department and trying to build (or at least share a building with) the new medical department that was being developed. We were on a quarter (rather than a semester) system, where normally the courses were four credit hours. David set up her courses so they would be five credit hours, with the extra credit hour being a statistical laboratory that operated just like a biology laboratory. This worked well because people didn't have their own pocket calculators back then.

Because of the extra credit hour, David essentially got more money from UCR than many other departments to teach the classes. The mathematics department in particular was very jealous of this. They didn't have labs, and so all of their classes were four credit hours. This was one of the strategic things that David did to get money for her department. In fact, there was a lot of contention between the mathematics faculty and David. They were jealous of the number of students, understandably. But they wanted her investigated for a variety of reasons.

David annoyed a lot of people, and in those days, it didn't help that she was a woman. But she needed to get things done!

One of the rules at that time at UCR was that teaching assistants should not be teaching the classes. So while David was teaching the introductory class in statistics, in the math department teaching assistants were teaching the introductory classes (calculus and so on), despite this rule. One professor who was in charge of all of these elementary mathematics classes had his name on the published schedule for 110 credit hours per quarter, which is clearly out of bounds. I went up to David while she was doodling in her office and said she should take it easy on him because he was teaching 110 units.

David looked up from her desk at me and gave me one of her stares. I thought I was going to melt. She got really excited about this information. She asked me if I had told Morris Garber about this. She seemed to tolerate me more after showing her that the mathematics department was abusing the system. It gave her some ammunition. That was my introduction to David. I survived one of her stares, which could be deadly if you got one of these in a lecture. With the reputation of the stares, people didn't go up to her very often. The stares were like deadly laser beams.

When she was starting the statistics department, David hired a variety of junior faculty members. Several of them were English (David Strauss and Chris Robertson). Two of them were spouses (Robert Beaver and Barbara Beaver). When we were having the war with the mathematics department, and because there were so many bad feelings, a group of us graduate students requested a meeting with the chair of the mathematics department. We asked "What's going on, what's the problem here?" Things were just very tense back then. Not that we thought there was anything specific we could accomplish. We just felt we had to do something.

A group of us met with the chair of the mathematics department. The chancellor of UCR also attended. One of the things we said was that David was only being allowed to hire junior faculty and no senior faculty. They said that wasn't correct. The chancellor said she'd been given every opportunity to hire senior faculty and didn't. We suspected she wanted junior faculty because she had more control over them and that with the senior faculty doing their own thing she had less control. In retrospect, it may be that she was just following in the British model where there was the one senior person (the professor), and they hired lecturers and maybe a reader.

Things calmed down a bit after the intervention. The mathematics department faculty wanted to start teaching statistics themselves because they saw that it brought in a lot of students, and since it was a requirement for many of the majors, it was a cash cow. That continued. But the demands from some of the mathematics faculty for investigations and things like that seemed to die down. The intervention meeting was probably 3 years before David left UCR—it was almost in the middle of her time there. I think the mathematics faculty were all men—she was fighting with all men.

In terms of physical stature, David was pretty short and round (she wasn't tall at all). In terms of personality, she was tough, but she joked around a lot too. Her humor included sarcasm or some physical humor. She went a little overboard at times. There was a very big football player in her classes. She talked to this guy often. One day we were walking down the hall, and for whatever reason, she took her keys and jabbed him in the back. This is not exactly humor, but she thought it was funny. The guy winced, and fortunately he was big enough to be able to take it and didn't complain, but we could tell she got him pretty good. She was playing around, but she had a pretty powerful punch.

Birds were a big thing for David when I was in the forest service. She had her "two-bang" theory for counting birds. Somewhat sarcastic on her part or maybe a whole lot sarcastic is that, if birds are in a tree, you'd take a shotgun and fire it to get the birds out of the tree, and then you'd fire it again to shoot the birds and count them once they were on the ground. Hence, the two-bang method.

On the serious side, she had a draft of a "statistics for the birds" manuscript that she never did publish, where she described some approaches for counting things and testing where there were trends in bird populations. David wasn't enthusiastic about the intelligence of people who counted birds, and she called them "bird brains." One of the things she kept telling us was that you wanted to learn something about the applied field but not too much so that you thought like them. That was really important to her: you needed to speak the language of whatever applied field it was, but you didn't want to know too much about it.

David didn't tell me much about her early days in England. She did tell me that Karl Pearson barely tolerated her at times. He would just look at her and shake his head because she was a woman. She told me that when she went on retreat with Neyman and other folks, Neyman and the other men wouldn't include the women very much. The men would go off and do work, and they would expect the women (including Elizabeth Scott) to do cleanup and other housekeeping chores. She wasn't too happy about this, but she recognized that there were certain limitations in those days, and there just wasn't much she could do about it.

David also told me that she had a bad motorcycle accident, which led her—instead of riding—to go fishing with William Sealy Gosset. The accident is why she used a cane a lot, including when she was working in Berkeley. She would cross the street, and rather than using her cane to walk, she used it to aggressively wave at drivers to make them stop. She had her fist around the middle of the cane and would shake it at them. She wouldn't look at the drivers, and she would just keep walking and shake the cane at them. When she was in Berkeley, she would mainly walk. She used to drive a Mini Cooper, but she had to give that up, and I don't know what happened to it.

David smoked cigarettes and cigars. I heard stories—I don't know if they are true—that David started smoking cigars when she was doing war work in England. She smoked cigarettes before that, but they apparently were really hard to get during the war. Cigars were a little easier to get.

At the forest service, David smoked both cigars and Turkish cigarettes, but mainly the Turkish cigarettes. We used to collect the tins and the cigar boxes at the forest service. I still have a few cigar boxes here and there. I found one of each in my garage (Fig. 10.1).

It was at a time when you could smoke in your office, but things were changing. The administration finally got her (and several others who smoked in the office) some air cleaners that were the size of large boom boxes. They would snap, crackle, and pop all of the time when they encountered smoke, which was pretty heavy in her office.

David had a good sense for how things worked politically in the University of California system. When the new building which would house the statistics department was being built, she requested extra space for all of the offices in the department. She wanted to give space to the students. There were certain size limitations given by the architects, but she demanded larger offices because she said we all needed to have punch card catalogs (i.e., cabinets) in them. Once she got the approval for the larger offices, she canceled all the card catalog orders. So everyone had really nice large offices in the new building.

In addition, David didn't want the campus folks moving anything; she wanted the faculty and graduate students to do it, so over a weekend we moved everything from the old building to the new building. It was a lot of work, but it saved money. David knew the ropes.

David essentially placed me in the forest service. She helped me get a contract with the forest service when I was a graduate research assistant. Then I was a teaching assistant, and then after I was awarded my PhD degree, I was offered a job in Hawaii (again, after David's recommendation), but I had to decline it because the grade level was too low (I had just gotten married and couldn't afford it). So I spent a year teaching at New Mexico State in Las Cruces followed by a year as a statistician

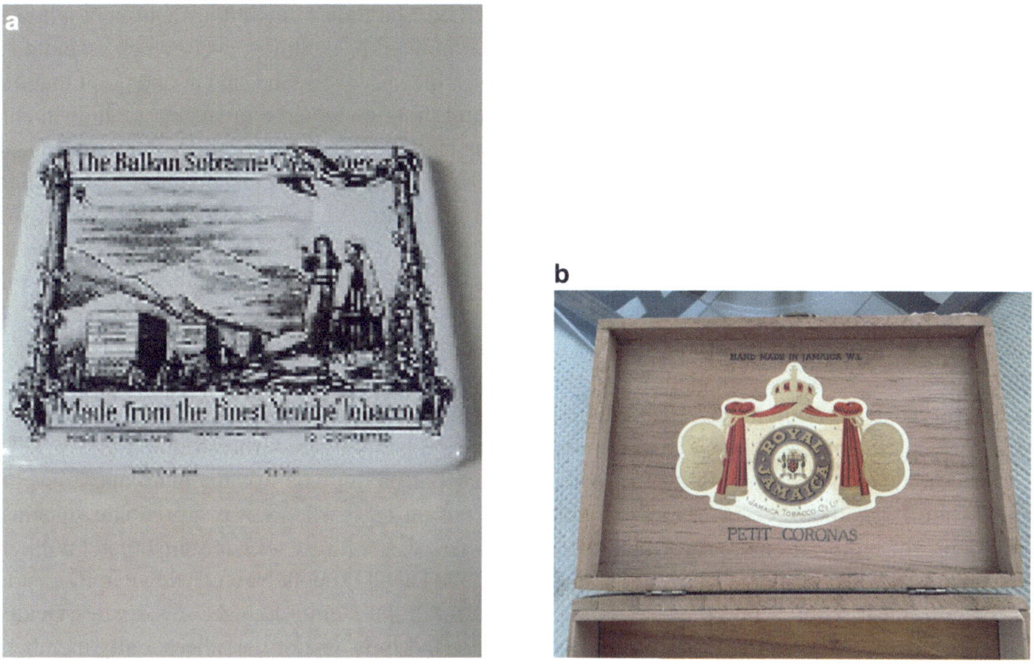


**Fig. 10.1** [David's tobacco product boxes saved for many years by Baldwin, J. (2022)] **(a)** One of David's cigarette boxes **(b)** One of David's cigar boxes

at Hanford, Washington (a nuclear waste facility). That meant I was qualified for a higher grade, and when they offered me the job again in Hawaii, I took it. I was in Hawaii for 2 years, and then Ronald Reagan became president and shrunk the size of the government. I was lucky that they were able to transfer me to Berkeley where David was, and I was able to keep a job.

The Pacific Southwest Forest and Range Experiment Station branch of the forest service had a fire lab in Riverside. When I was a graduate research assistant at UCR, I was on a contract that David arranged with them to map data on forest fires. They had all these sheets of paper where people wrote down where the perimeter of the fire was, where it started, what it was on this date, and what it was on that date. We entered all of those data into a database.

The forest service was a long-term client of David's. She consulted for them even before going to UCR, and then the whole time she was at UCR. So David was already working for the forest service when she was on sabbatical in Berkeley. She commuted almost once a week going back and forth flying from Oakland to Ontario to get from UCB to UCR. One of the faculty picked her up and dropped her off at the Ontario Airport. Usually Vahram Sevacherian did the chauffeuring. She'd be in Riverside during the week and then go back to Berkeley on the weekends. She did a lot of commuting. I would guess that David got hired by the forest service through Neyman at UCB.

David Sharpnack was David's supervisor and my boss when I worked for the forest service early on. Joyce Dye was one of the clerical people in the forest service. In David's last days, Dye took David to medical appointments and cared for her almost on a daily basis.

I don't believe David ever became a citizen. In fact, she bitterly complained about a tax issue and how large a percentage of her income went to taxes. Because of this, when she was at the forest service, she had to be fired every year and rehired. She regretted it many times that she emigrated to the United States because of taxes.

David lived in an apartment in Riverside while at UCR. She didn't buy a house but kept a house in Kensington near Berkeley (which she owned) and commuted back almost every week. She didn't make the adjustment from big city London to rural Riverside very well. She said at one point that the biggest mistake of her life was going to Riverside. I don't think there was a great deal of adjustment. It doesn't surprise me that she sent her body back to the United Kingdom for burial.

David doodled a lot when listening to lectures. She had a scratch pad of just doodles. She doodled all the time, drawing generally geometric figures over and over. When she listened to lectures, she would rarely look up—she would be doodling constantly.

If someone ever brought up a Bayesian argument of some kind, David would walk out of the room. She didn't seem to tolerate Bayesian thinking very much. So my Bayesian education early on was not the greatest. Jim Press took over for her as chair when she left. Given that he was clearly a Bayesian, I don't think they got along very well in the transition. I don't know David's exact reasons for leaving UCR. She might have just gotten tired of the position and the commuting every week. Also, the department had gotten up and running, and she probably figured it was time.

David invited a few people out to Riverside. She had Barton out—he was the main one—several times while I was there. Colin Mallows didn't come out (at least when I was a graduate student), and I don't remember if Norman Johnson came out. There was a forest service statistician (William O'Reagan) who was retiring when I was a grad student, and David brought Neyman down to Riverside for the retirement party. While there, Neyman drank a lot at the faculty lounge—a statistics faculty member said he had about ten martinis. Because of those martinis, one of the other faculty members and I had to keep Neyman balanced for a while and walk him to the table where we had dinner. We sat him down, and then every 15–20 minutes, he would sort of stand up from his chair, raise his glass, and say "To the Duchess!" And then he would sink back down to his chair. He did that several times. We never got an explanation as to why he called David that, but it was clearly something he had called her before.[1]

I computed lots of statistical tables for David while we worked together at the forest service. We did have access to a computer back then—not great computers, but I was able to use them to do tables and moments for her. These were a lot like the things she did for the correlation coefficient that she did by hand. It is just amazing how few mistakes were in the tables that she did essentially by hand.

A year or two before she died, some of the programs that did symbolic algebra like Mathematica came out, and I really wish I would have had the time to use it, because you could do things far exceeding what she could possibly do by hand. I just never got the chance. My biggest regret about working with her is not being able to help her more. I think of how she would have liked the ability to extend the combinatorial problems she worked on.

## 10.3 Roxy Peck Presentation (1 August 2019)

These words describe David: feisty, unconventional, persistent, uncompromising, independent, prolific, generous, and charming.

I had been given a bachelor's degree in a social science from UCR, and I decided I wanted to go back for a second bachelor's degree, in statistics. At that time, many people were trying to go back to school in order to avoid the draft, and a lot of paperwork was required to be admitted to a bachelor's degree program. So the UCR statistics department suggested that I enroll in their PhD program in statistics. There would be less paperwork, I could take all of the undergraduate courses in my first year of the PhD program, and then I could petition for the bachelor's degree and go on my way.

[1] Note: Neyman called David *Duchess* because she was stubborn and wasn't afraid to disagree with him.

I proceeded with the plan, but when I reached the end of my first year, David told me that even though I met their conditions, they weren't going to give me a second bachelor's degree. Instead, I should stay and complete the PhD degree. David also said she would take care of my support: it wouldn't do for me to continue to support myself by waitressing. David was creative at recruiting and supporting students.

At the graduate student orientation, David explained that her department was housed in a new building. The computer center was on the first floor where there were only men's restrooms, and the statistics department was on the second floor where there were only women's restrooms. She told us of her desire as a young woman to become an actuary, but when she showed up for the interview, she was told they didn't hire women because they had no restrooms for them. So in her new building at UCR, women were accommodated on the statistics floor, and men had to go downstairs to the first floor to use a restroom. David was gutsy.

My first statistics course in the PhD program was David's course on Games, Gods and Gambling. There were no prerequisites, which followed from the fact that the corresponding landmark book, which she wrote, was aimed at a lay audience. The course covered the development of probability theory during the past three millennia, which is a lot to cover in only 2 hours a week over 10 weeks. She talked about how dice used to be made out of bones, and the religious leaders would predict the future by rolling the bones and seeing how they would land. The course was fascinating, in large part because David herself found the subject exciting.

My last statistics course in the PhD program was David's seminar on special topics in the literature. David required every graduate student to take this course before advancing to candidacy. She would assign students a topic, and then she would let the students select a topic. In both cases, she required the students to give presentations on the topic. She would critique, often harshly, and then she would go around the room and require each of the other students to say something critical. It was a totally traumatic course.

But, when I had to go out to interview for academic positions, I was ready! David's methods were "tough love," but I really came to appreciate them. When I later expressed my appreciation to her, David replied, "Oh, that's very nice, dear, but really, I was just making sure you wouldn't embarrass me!"

In the conversation article about David, it mentioned she was chair of the university affirmative action committee. In her tough love mode, she would say that women would come in crying and left in tears because she wasn't very sympathetic. But then she'd pick up the phone and call up the guy and say "I'll take you on before the academic senate [2]."

I became aware as a student that David wasn't a fan of Bayesian statistics. In fact, she was very anti-Bayesian, and we knew we shouldn't ask anything about it. One time she took her graduate students to a conference, and when someone started to give a Bayesian talk, she shook her cane and left the room while knocking over her chair as she walked out. As a result, I first learned about Bayesian statistics after I completed my PhD degree and was in my first teaching job. When David left UCR, they hired a Bayesian to replace her. David had hired all of the faculty in the department up to that point, which made for an interesting transition.

We also knew we shouldn't ask anything about factor analysis, because David did not approve of that method. I learned this at my oral exams, when the outside member on my committee (someone from the sociology department) asked a question about it. David flew out of her chair, refused to let me answer (even though I could have done so easily), gave a 10-minute lecture on the evils of factor analysis, and then sat down and said "Next question?" Needless to say, the outside committee member didn't say another word for the rest of my exam and probably has never agreed to be on another statistics committee!

You will hear it said about Dr. David that she won the Elizabeth L. Scott Award and is described as a role model, mentor, and teacher. I have to say she was all of those, but she wasn't any of those in a typical sense.

David was a role model, but not a typical one. I don't think any of us aspired to be like Dr. David. She could be a bit cantankerous and feisty and smoked cigars. None of us actually pictured our future to be like her. And we also knew that we weren't in her league in terms of statistical abilities. But the thing about David is that she allowed you to see the possibilities of what you could accomplish if you worked hard. And what she valued was people who would work hard for what they wanted.

Likewise, David was a mentor, but not a typical one. I would describe her as part of the tough love school of mentoring. In her acceptance speech for her Elizabeth Scott Award, she said that, although women had come a long way, it was still important to teach the young that when you meet a set-back, don't sit down and weep, just go on with your work. That was her philosophy. Even though she wasn't your typical mentor, she actually had an interesting way of mentoring. I have a couple of stories.

One story is from my second year of graduate school. I was starting to wonder if I made the right decision to stay in the PhD program. David took a group of students to the WNAR[2] meeting, which was in Santa Barbara that year. I got there and looked around the room: it was mostly old men (two of whom had socks on that didn't match) who were pretty much ignoring all the women and everybody under 30 (not that there were many of either of those in the room). I was really starting to think: I'm not sure there is a place for me in this profession.

And then on the second day of the conference—David had us sitting in a row like we were her ducks, and she was sitting in front of us, checking on us periodically—the next speaker was a very young Jessica Utts.[3] She was presenting on her master's work and began talking about a goodness-of-fit test she had done, which she called the rainbow test. You can imagine the reaction from this audience when she was introduced as a speaker, and this title was given. You could just see the eyes roll back in the heads of the people in the audience. But Jessica got up and stood in front of that group, and she just smiled and started talking. And slowly you could see some of them starting to listen and pay attention.

David turned around and looked straight at me and she said "See!" And it was really her message to say "There is a place for people like Jessica, for people like you." That was a turning point for me. It made me realize that I should stay and continue to work on my PhD degree.

The second story is from many years later, after I left Riverside and was on the faculty at California Polytechnic State University. The director of student services there and I had gotten to be friends. He was injured in an accident when he was a young man, and he ended up in a wheelchair as a paraplegic. One day while we were having lunch, we discovered that we had been at UCR around the same time. He said to me "Did you know F.N. David? and told me a story."

The story was that he had just gotten back to school at UCR in his wheelchair, and he was still angry and feeling sorry for himself about his accident. One day he was on campus and wheeling up to the door of a building when David put her cane in front of him to stop him, go around him, and open the door for him. In his angry mindset, my friend snapped back at David and said "I can do it myself." She turned around, blocked his way through the door, and looked at him and said "I would have done the same for anybody, get over it, and get on with your life." Then she opened the door for him to go through. It was interesting because he said "that was what I needed to hear at that moment in my life." That made such a huge difference to him. David never hesitated to give personal as well as professional advice.

[2] Note: WNAR is the Western North American Region of the International Biometric Society.

[3] Note: Jessica Utts went on to become the 2016 President of the American Statistical Association.

David was a teacher, but also not a typical one. Actually, I think in some ways she was a legendary or star teacher because she was ahead of her time in incorporating what we today consider good practices. She believed in teaching statistics by having students work with data. She used to put herself in the students' place and play out the discovery process with them on how to do statistics. She would walk around the class while talking continuously and verbalizing things that were going through her mind. When she went around the room talking, you got to hear her thinking out loud and express the thought processes. Then she would turn around and expect you to do the same thing. For her time, she was a very atypical teacher.

But there was an elephant in the room: David told colleagues that she loathed teaching. For example, she said that her classes got so large, about 450 students when she taught her history class at UCB, that things would get out of hand. As a student, I didn't know that David disliked teaching, because she was very popular as a teacher. I discovered her dislike of teaching much later. From a student's perspective, this situation was interesting. David never let her dislike of teaching show in the classroom. I would say that, while she may not have liked teaching, I'm convinced that she really did like and care about students.

At the bottom line, what David really wanted to do was research. She viewed teaching as something that she had to do in order to do her research. When she retired, she happily gave up teaching and continued with her research.

In summary, David was not your typical role model, she was not your typical mentor, and she was not your typical teacher. She was all of those things (role model, mentor, teacher), but she wasn't typical. And it's not surprising because there was absolutely nothing typical about her.

David fought a lot of political battles at UCR. The establishment of the statistics department there was very controversial, and she especially fought tooth and nail with the mathematics department. In general, she took on anyone who got in her way, because she knew she would have to fight for what she wanted.

Apparently David was very successful in these political battles and came out on top. She got what she wanted, although there was blood all over the floor. For example, she talked a little bit about not being able to hire who she wanted in the hiring process: The university said she had to go through a search, but then the people she wanted were hired.

I think David would have loved being called a mathematical badass. Her demeanor was rough, but she also had a gentle side in the sense that you really thought she cared about you and what you were doing. On the gentle side, I have a memory. The building that housed the statistics department had just been opened. They had a dedication for the building. Neyman came down from Berkeley to talk at the opening. It was interesting to watch their interaction: Neyman kept toasting her: "To the Duchess! You could see that the two had a very special relationship."

David didn't treat women and men students differently. She was just pro hard work. She didn't see those other things. She expected people to work hard for what they wanted. That's where I saw some differences in how she treated people. She really valued people who would work hard.

## 10.4 Richard Green Interview (27 July 2020)

I met David sometime in the mid- to late 1960s. It was when I was a graduate student in the statistics department at UCB. I was a student of Jerzy Neyman, and I was one of his people. David was one of Neyman and Elizabeth Scott's friends, and I would see David at parties, dinners, and other get-togethers in Berkeley.

David said that, when she was little, she told her parents she wanted to be called F.N. She applied to be an actuary, and she must have done it under the name F.N. David. They asked her for an interview, but then when they found out she was a woman, they said "You can't have a job here because we don't have a woman's job." Actuaries were not supposed to be women. I think people called her F.N. People at UCB always referred to everybody as Mr. and Miss, like Mr. Neyman and Miss Scott, and so I always called her Miss David. Probably other people called her F.N., but I'm not sure if it was to her face. One person called her Flo. I assumed that was "to get her going."

I was a graduate student at UCB for 9 years and didn't complete my degree until the 10th year when I was teaching at UCR in David's department. Some time before the summer of 1970, I took David's History of statistics class at UCB. I remember her to be a very good teacher. I recall David's comment in the class about Pascal's triangle. She said that it was named for him because he was the last person to discover it. This was some time before "Stigler's law of eponymy." David thought that Fermat was the greatest mathematician who ever lived. Fermat was certainly a great mathematician, but I think that part of her thinking about him as a mathematician was influenced by the fact that she admired him as a person.

I kept my apartment in Berkeley for several years after completing my PhD degree and while I was working at UCR. David kept a house in Berkeley (Kensington, actually) and a badly furnished apartment in Riverside. Quite a few graduate students and some young faculty members lived in the same apartment complex. Because she and I likewise had residences in both Riverside and Berkeley, I traveled between Riverside and Berkeley a number of times with David, particularly during my first year at UCR.

David also kept a car in Berkeley, which she would leave at the airport. She once commented that it would be cheaper to buy a house in Riverside than to park each week at the airport. But she had good friends in Berkeley, particularly Neyman, Scott, and Lucien LeCam. Perhaps her best friends in Berkeley were LeCam and his family. I don't think that she liked his too-theoretical work, but she liked him personally, and she admired his physical strength. There was a saying in Berkeley: If you say something is "known," it means that LeCam knows it, and if you say that something is "well-known," it means that someone else knows it too. I think that LeCam was the most open of the professors in statistics at UCB when I was there. He was willing to talk to anyone, and he seemed to not worry about wasting his time. He knew a lot, and he was not just interested in his own ideas.

Riverside had been the site of the citrus experiment station. After the founding of UCR as an agricultural campus in 1954, there were moves to make it a more general campus. Morris Garber, who did much of the work of starting the biostatistics department (which became the statistics department), and David were the only two statistics professors at first. David started as statistics department chair at UCR in 1967.

I got to UCR in 1972, 5 years after David arrived as chair. David asked me if I wanted a job at UCR. I was not sure and churlishly did not answer. Some time later she asked if I wanted a 1-year job. I agreed. She brought me down for an interview with the dean, Mack Dugger. It was a funny interview, mainly with David telling the dean what was what. In those days, at least at UCR, job announcements were by word of mouth, and job talks were not required. A number of oral histories testify to this. My temporary position turned into a permanent position, or it would have been permanent if I had gotten tenure, which I did not get, but I did stay at UCR from 1972 to 1980.

When I first went down to Riverside for my job "interview," I had a long talk with David. I was surprised at how critical she was of the work of her friends. She thought, for example, that Neyman's work on bombing for the World War II effort was wildly wrong because he believed what he was told by military people, who grossly over-rated the accuracy of bombing. I am quite comfortable with this sort of hyper-critical attitude, since I am both critical and not too attached to my ideas, but I think that some of David's UCR colleagues were a bit thin-skinned. There is an old saying—I saw it made of Eleanor Roosevelt—that "She knew all about us, and she liked us anyway." I think that may apply to David. Some people probably prefer unqualified admiration.

I felt that David had a good impression of me, and I didn't know why. As a junior faculty member, David supported me by giving me a quarter free of teaching so that I could finish my dissertation. She wasn't on my dissertation committee at UCB, but I believe she did some arm twisting with her colleagues there to get my dissertation approved. When I later moved to a faculty position at the University of Minnesota—Duluth, I was up for tenure, which I got after she wrote me a nice letter of recommendation. I certainly liked her and think she probably liked me, too.

This was my impression of her as a boss: She would tell people what to do, and she would ball people out, and she'd say "Pull up your socks!" But if someone complained about the department, she wouldn't transfer the pressure. In my experience, if someone is under pressure, they will pass it on. But not David. If someone complained about the statistics department, she would come back and say "You have no business telling us that!" My idea is that a good boss doesn't yield to outside pressure, and she didn't seem to do that. She was tough.

David had presence. As a department head, David had to be tough. She was a busy department head and stopped publishing sometime while she was at Riverside. She thought since she wasn't publishing anymore she should teach extra classes, so I think she was teaching three classes at a time, when the usual load of courses across three quarters was 2, 2, and 1. But she didn't like teaching, which was sort of funny because she was very good at it. David taught students at many levels. She advised graduate students, but she also explained arithmetic to students who did not understand it. She once told a story of a UCR student who did not understand subtraction and negative numbers. She took the student out in the hallway and took some steps forward and added a negative number by taking steps backward.

The UCR statistics department and computer center moved into a new building a couple of years after I got there. David used to talk about *turning the crank*, which is something she had done a lot of in her early days when she used manual calculating machines. Her time at UCL was before computer science. You couldn't do then what you can do now with high-speed computers. When she was at UCR, statisticians used computers, but mathematicians did not. Before around 1974, you would prepare a stack of punch cards and bring them to the computer center in the basement of the building that statistics was housed in. When we moved into the new building, we had ten professors (David, me, Bob Beaver, Morris Garber, D.V. Gokhale, Sue Johnson, Chris Robertson, Vahram Sevacherian, David Strauss, and Larry Zahn). I think that half of the professors were hired in 1970 and 1971.

David certainly did applied work during the war. That was one of the things she talked about. She did theory—she did symmetric functions—but I think of her as doing applied work. She worked with Frank Vasek in the biology department. He was interested in creosote which grows as bushes in the desert. Because the plants repel each other, they don't live very near each other. If you look at the plants in the desert, you see a ring of plants and then another ring of plants. You start with a bush, it repels other ones, and then eventually it stops and dies. She was interested in that sort of distribution of creosote plants. She had the maps, and when she would fly back and forth between Riverside and Berkeley, she'd be looking at the pictures and thinking about them. She would say "Live with your data."

In my estimation, David was really interested in applied, practical problems. She would also do some theoretical work when there weren't already appropriate methods for dealing with the real problems. She did things she and other people then applied to the real problems. For example, she worked on real problems that had both space and time components and developed mathematical theories that epidemiologists probably would not think of. In fact, David's department was really useful as a bridge between academics and people trained in applications.

One of the important points was that the department at UCR really was about the fields in which statistics was applied. This was different from the purely theoretical work that people did at UCB, which they were really good at, but which often was really theory of applications.

Beaver, I think, had a joint appointment. What was happening in UCR is that people had problems, e.g., agricultural problems, and they wanted the statistics department there to help them. Sevacherian basically had a job helping people from entomology with statistics. I didn't know computers, and there were people at UCR who helped me write programs when I needed to do calculations. You had people who were really happy to have work to do. That was one of the great things about UCR. There was really good support.

Robertson was someone David was not comfortable with. He was also from England and had been at Cambridge. David was, I suppose, upper middle class, and she was pretty good at treating people with condescension. She thought that Robertson's parents were pretentious, and he thought that she was condescending to them. So there was certainly some not very good feeling there. My idea was that she would tell people "you should be better or something." She would criticize you and also descend you. But there were people who had higher values for themselves than they should have, and that caused irritation.

Sevacherian was to David sort of like what Scott was to Neyman where Scott was the driver/helper in many ways. Sevacherian also was a student at UCR, so he knew David probably from the beginning. He was Armenian, and his family had a business in Los Angeles. He was teaching at UCR in the statistics department. He was an entomologist, and he was also helping his family with their business. He was probably the person who knew David the best in Riverside.

Strauss was also from England and had been at Cambridge. He was someone David admired because he had a first-class honors degree from Cambridge, and he was a chess master. He had just won the US Open Chess Tournament, which is a national, serious chess tournament, the year before I got there. He won that several times. He played chess when he was in school, but Riverside had somebody who was a really good chess teacher. . . . Strauss probably improved his chess in Riverside. Ross Stoutenborough, who was also a master, was a student in the department.[4] And then Larry Christensen, who was maybe the best player in the country, was also at Riverside. Strauss taught a class for a while on game theory, but it was really chess. I took the class. You could go to the class and play chess. He talked about chess, but he also played chess with the class.

I knew David very well. I certainly spent a lot of time with her and saw her a lot. I traveled with her and went to dinner with her once a week when she and Sevacherian and I had dinner together. Maybe we were not friends, but certainly I was friendly with her. But I didn't know much about her hobbies, except that she did jigsaw/crossword puzzles, and I was never in her house in Kensington (I always thought she lived in Berkeley).

Floyd Maxwell was a black guy. He and his wife (Roberta Runnestrand) both got their degrees with David as their advisor at UCR in 1974. David would call everybody "Dear boy," including Maxwell. My idea about her was that people regarded her as eccentric, so she could do things that were natural for her, and people might think it was strange, but I think that people didn't take offense because they thought "it's not me, it's her."

David dressed just like a man when I knew her at UCR. I'd say she was buxom, and she would have a load of stuff in her front pockets. She wore shirts and skirts. I don't remember the details, but she would carry a pack sometimes. Now kids carry packs, but people didn't in her day. She smoked cigars. You could say that she acted like a man.

My idea was: I liked her very much, and I could describe her as kind and gentle and motherly, but if I were to give a phrase for her, I'd say "she was a tough old bird." She wanted people to think of her as a tough old bird. One time I slammed my hand in a locker or something. I was bleeding. I went to her like I'd probably go to my mother. I went to her apartment and asked her for bandages and

[4] Note: Stoutenborough became a master only 3 years after he began playing chess. He became a computer programmer at UCR and directed the Riverside chess club.

sympathy. So I thought of her, I suppose, differently from her colleagues who were offended by her rough treatment sometimes.

I didn't know anything about David's family or personal relationships. She did mention something about having a boyfriend who was a pilot and who was killed. That would have in the 30s I guess. She didn't give any details. As far as we could tell, David was friends with Scott. Certainly with David, we were interested in so many things like science and what was going on in the world. There was a lot of that kind of thing to talk about, and so we didn't talk about personal things. She had some rude things to say about many people. But I think she was on reasonable terms with those people.

David had a friend named Joyce who stayed with her for a while in Riverside. David and Joyce planned to live together after David retired. But Joyce died. David gave the impression that it was something that Joyce should not have done: Bad on her part. Neyman had a heart attack and then had another heart attack and died. David said he knew he wasn't going to recover and just decided to die. David was tough with herself and tough with other people.

At the same time, David was really helpful to people. When a new person would come, and if someone was buying a house, David would arrange for everyone to go out and help. Another thing she would do is have dinners with faculty and graduate students. Mondays she'd have dinner with everyone, Tuesdays she'd have dinner with Sevacherian and me, and then sometimes she went out on Thursdays. She'd spend her weekends in the Bay Area. I think she'd get a roast and cook it on the weekend, and she'd bring fixings for lunch back with her on Sunday.

David and Sevacherian would help people. They took me shopping for stuff for my apartment. I wanted to get two pillows. One of them was two for 6 dollars, and one was two for 7 dollars. I didn't know which one I wanted to get, so I asked to get one of each. Miss David said "Can he get one of each and pay 6.50 dollars?" The salesgirl said "No." And I said "Can I get one of the 2 for 6 dollars and pay 3 dollars and one of the 2 for 7 dollars and pay 3.50 dollars?" And the salesgirl said "Yes." And David said "That's what I said." And the girl said "Well, I didn't understand your accent." And David said "I don't have an accent. You have an accent."

I think David was like that! She would say what she thought. And people would take it as: This is some strange woman! Rather than: This is someone insulting me, or confusing me, or something.

There was a rule that one had to retire after the academic year in which one reached age 67, so David was required to retire in 1977. I am not sure, but I think that David did not want to retire, or if she had to retire, she wanted to be "recalled" to active service. Neyman had retired before I met him, but he was "recalled" every year until he died. It seemed to me that David alienated or angered some of the faculty. Whatever the politics, the last year of David's time at UCR was unpleasant. I don't know what the details were, but she was sort of mad at everybody that year. One of the things was with Garber, the guy who had started the department and had been there from the 1950s I think. He started out in biometry. He published a huge number of papers. People would ask him for statistical advice, and he'd want to be on the paper. David didn't like that. She thought that, if you advise someone about statistics, they can acknowledge your help; you certainly shouldn't insist on being on the paper. If you do stuff that's original statistics, then you publish it in a statistics journal. She thought, basically, the stuff that he got credit for (publishing or having your name on papers) wasn't very good. So I think there was a bad feeling because of that.

Also in terms of politics, the mathematics department didn't like David. David once told a story about being on a mathematics department committee for a faculty member who was going up for promotion to full professor. She was the only one who bothered to read the candidate's papers, and, based on what she read, she disagreed with the department that he should be promoted.

The mathematics department also didn't like David's department. The two departments quarreled, with a result that, if you wanted to be a statistics doctoral student, at first you had to pursue a PhD degree in population biology; only eventually, could a student pursue a PhD degree in biostatistics. As I understand it, mathematics would not permit the PhD in statistics. I think the PhD in population biology, which statistics students could receive at first, was a work-around. When I got there in 1972, there was a population biology group with members from a number of departments including statistics, entomology, biology, anthropology, geography, and nematology (very few universities had/have nematology departments). As long as David was at UCR, there was hostility between mathematics and statistics. Decisions were made, at least in part, by the academic senate. David would often "call out the troops," that is, get her many friends to go to senate meetings to vote for the statistics side.

As a senior person, David knew everybody and was friends with a lot of people. She was comfortable with senior people. For example, David knew William Sealy Gosset and was friends with him. She had a fairly serious accident on a motorcycle, and he was really friendly with her when she was recovering. She remarked about his being eccentric: He liked to be in a rowboat or canoe using an umbrella as a sail. David referred to herself as being Neyman's first student. She knew and worked with Neyman and Pearson. She said that Pearson was a good teacher, but she learned more from Fisher than Pearson. Fisher did not acknowledge her existence and would not let her ask questions. For another example, she knew J.B.S. Haldane. He was one of the people everybody was afraid of, but she was not. She was certainly brave. When she went to the dentist, she did not want to have any anesthetic because she wanted to know what that dentist was doing.

David immigrated to the United States because she was a woman; she wanted to take E Pearson's position when he retired, but she was not given it. She was professionally disappointed in that. There was a picture of her at UCB in 1960 before she left the United Kingdom, so she must have been coming to UCB. I am sure Neyman would welcome her, and she probably knew UCB was a place that she could come. I think she was disappointed in UCR. She was unhappy in her last year there, and she made everybody else unhappy too.

When David retired, there were three untenured professors in the statistics department at UCR (Johnson, Sevacherian, and me). None of us got tenure. The deciders thought they could do better by replacing us. Johnson had not published many papers. Sevecherian went to entomology, where he did get tenure. He was basically an entomologist anyway. I think he wrote a thesis on estimating insect population size, so he was doing statistics. He was good at what he did, and he was really good at what the department had been before, which was helping people with entomology and pest management. I went to the University of Minnesota—Duluth department of mathematics and statistics. I retired from there in 2018.

While student numbers increased at UCR, the number of professors in the statistics department did not for some time reach the number that it had when David left. I met Barry Arnold at a national statistics meeting in 1986, and he said that the department was still seeking authorization to fill the "Green position." I am not sure what happened, but my guess is that the UCR statistics department was begun, at least in part or perhaps largely, to provide statistical help to agricultural experimenters. Beaver's appointment, for example, was two-thirds statistics and one-third experiment station. The department had a number of technicians who worked with agriculture.

I think that David's department was very good at what it was originally intended for. There were people, both faculty and staff, who were very good at helping people with statistics. The secretaries were very efficient. I operated on the principle that if I wanted a long letter typed and ready for the 3:00 mail, I had to give it to the secretaries by 2:00. Gokhale mentioned after a visit to a major university (Ohio State, I think) how much more efficient our department was. My guess is that after David left (and Jim Press came in as head) there was an effort to upgrade the statistics department quality by getting people who were better at mathematics and theoretical statistics and to move away from helping researchers with their statistics.

The lesson I learned was that, if you want to be an applied statistician, you shouldn't work at an academic department, because what you are good at is doing statistics. That's not what academic departments want; they want you to publish papers in the *Annals* or the *Journal of the American Statistical Association* (JASA) or wherever. If you were at UCB and you published in the *Annals,* that was good; if you published in JASA, that was ok; and anywhere else they would ask, why are you publishing there?

With our departure, and with the departure of David, I think the department repositioned and became a more standard statistics department. It became smaller, and more women were recruited. They did have a problem hiring people, but my guess is that they eventually got good people.

## 10.5 Sander Greenland and Peter Lansley Comments (27 May and 23 June 2021)

Sander Greenland is professor emeritus of epidemiology and statistics at the University of California—Los Angeles.

Dear Amanda,

I just read your article on Elizabeth Scott and F.N. David and wanted to thank you for it [4].

In the early 1970s while I was at UC Berkeley in grad math with a concentration in stat, Scott was my stat advisor, and I was seconded to David as an RA by Neyman when she was visiting there.

They were both very supportive, and I have fond memories of them (especially of David sheltering me from Neyman's ire when I took a Fisherian view of the problem I was assigned).

Having a shared interest in astronomy, I knew some parts of Scott's travails as a woman in that field and appreciate the further details and perspective you supplied. Also, I did not know about David's glass ceiling experience in the United Kingdom that led to her becoming the founding UCR chair—I always wondered why she made that move (she tried to recruit me there, but I chose UCLA instead which was much closer to my parents and had the health science concentration I wanted; UCR did not have any health science way back then).

Thanks again!
Sander Greenland

- - - - - - - -

Peter Lansley is professor emeritus of construction management at the University of Reading. The queen awarded him the MBE honor in 2014 for his work on day-to-day improvements for the elderly [5].

Dear Professor Golbeck,

Thank you for your paper in the June edition of *Significance* magazine [4]. It was fascinating and thought provoking and has led to a trip-down memory lane which you may find interesting.

. . . my comments below are about F.N. David. As soon as I knew her full name, I was fascinated by her, how couldn't I be. She was named after one of my childhood heroes!

. . . From 1965 to 1968, I read for the mathematics BSc at the University of Leicester and, in the final year, took a statistics option which was largely a probability course led by Dr. Rahman. He was a studious chap, who followed a very traditional syllabus and, importantly, was keen to help students choose the right career. Perhaps because it was an introductory course, most of his examples were taken from the work of legends of probability, Bernoulli, Laplace, and so on, as well as folk like Fisher and "Student," but every so often there would be examples from the work of F.N. David although there were very few from other contemporary statisticians. Judging by his approach, and his asides, he held F.N. David in very high regard—a living legend perhaps.

Rahman persuaded me to apply for a place on the masters course at the University of Newcastle at which the professor of statistics was Robin Plackett (b.1920–d.2009).... Plackett's lectures would occasionally refer to the work of F.N. David and so too those given by his colleagues. Sometime during the course, Plackett gave a lecture to the master's degree students on the "great and good" in statistics at the time: C.R. Rao, Henry Scheffe, John Tukey, and others from the United States and of course home-based ones (although he didn't have a lot of time for Lindley and the Bayesians) and he gave prominence to the importance of the work of F.N. David. However he added that her move to the University of California was a disaster (she was moving at about that time). At first I thought that he had implied she had made a poor career move, but later when I understood what she was achieving and knew more about the statistics scene, I realized that he meant that her permanent move was a disaster for statistics in the United Kingdom, especially for him, and your paper confirms this. Unlike today this was when international collaboration was an extremely slow process and quite unusual.

After a few years working as a biometrician, I moved to a private business school to join a research team working on the economics, finances, and organization of the UK construction industry. The team was led by a very powerful recently retired civil servant Eleanor Lea (nee Florence Eleanor Channon James (b.1906–d.1985) known as Lady Lea as a result of her husband receiving a knighthood a few years earlier. Eleanor was the first woman to study an undergraduate course in statistics at UCL. She moved into materials and other research relating to the construction industry in a technical division of the Ministry of Public Building and Works, but when she wed Frederick Lea, a renowned concrete chemist, who worked in the same division and who later ran the UK's building research station, she was obliged to work somewhere else in the MPBW. It wasn't appropriate to work in the same division as a spouse! This was about 1939. Eleanor went to the construction statistics division, developing methods for measuring construction activity during the war and later the rebuilding of Britain after the war. This was vital work informing government policy and action. So Eleanor became more of a business economics cum statistician (some of us would call her a "Statistician of the Blue Book—The UK National Accounts"). She retired from MPBW when it was reorganized into the department of environment (DoE).

An example of how straightforward and perhaps intimidating she could be was when she asked me to comment on a draft of an article she had written. I replied that it was very good and made some other complimentary remarks. She was not pleased. "When I ask someone to comment on a draft there are no 'Queensberry Rules'." (The rules of boxing!)

When I used to visit DoE, nearly everyone would say "so you know Lady Lea, I bet she works you hard!" They would also add "did you know that when she left they had to replace her with three men!" Yes, I did know, and I knew the three men, who always beamed when her name was mentioned and were immensely proud to have followed in one-third of her footsteps! So Eleanor was one of those few women who did break through the glass ceiling of the civil service.

Perhaps rather strangely in the 1970s through the 1980s, nearly all the leading statisticians in research involved with construction economic and policy work were women. Not all had high-powered jobs, but men were conspicuous by their absence!

I haven't yet mentioned F.N. David in relation to Eleanor, but that is next, in a rather round-about way.

Almost as soon as I started work as a statistician, I became a fellow of the RSS (about 1970). In those days a new fellow just had to prove that they had a keen interest in statistics and have two credible sponsors. Then I attended the meetings of the RSS, which were very formal indeed. Old-fashioned raked lecture theaters, with the speaker almost in a bear-pit. Galley proofs could be collected at the door of the lecture theater, and most presenters just read their papers, which could be very tedious. These would then be published in the RSS journal. But when the speakers had finished, the fun began, with the proposal of a vote of thanks, always by a very well-respected figure. After a few polite words, the proposer would then proceed to highlight some of the shortcomings of the paper some times in not-the-most-friendly way but would then thank the speaker anyway. The seconder of the vote of thanks, usually an up-and-coming figure, would follow the same pattern but usually without any boxing gloves on. It could be brutal. Then the seconder would say a few nice words, and the vote would be passed with acclimation. If the speaker was an old-hand, he (rarely a she) would be well prepared and would punch back and put the seconder in his place. Maurice Kendall seemed to relish the fight. However, those unfamiliar with the routine could be reduced to gibbering wrecks who could only say "thank you for those interesting points, I'll have to look at them."

So taken aback were some of us by the impressive way, Vic Barnett took a paper apart that afterward we had to go to the pub round the corner for a drink to recover! It was rather strange that so many of the senior folk at RSS meetings who came over as very formal and difficult were different people when they attended meetings of the Biometrics Society, where there were very friendly and supportive toward each other and especially toward young statisticians. Frank Yates, John Jeffers, and David Cox I recall as being particularly nice people.

When I related these experiences to Eleanor, she said that I had seen nothing much compared to what RSS meetings were like when she was a student, and she very much blamed an ill-tempered R.A. Fisher for establishing the tone of meetings. Apparently Fisher was a very combative character, who was known to enjoy bearing a grudge. He hated Karl Pearson and Egon. "It was awful, their studies looked across the quadrangle at UCL towards each other, and Fisher would stand at his window for ages (probably in a cloud of tobacco smoke), with Pearson at the other, scowling at each other. If ever Pearson gave an important lecture, Fisher would be sure to be there, and would always make sarcastic comments, and at times denigrate Pearson's minions who were calculating his statistical tables."

The suggestions from this conversation are that the atmosphere at UCL must have been very difficult for some, one of suspicion and fear of attack. It is this which F.N. David would have encountered when she joined UCL—but her work came first, and she found a way of working with the Pearsons (although at times she would have had to be very assertive), but ultimately she was let down by them. . . .

Apologies for this rambling message, but your paper reminded me that while F.N. David may have been an under-appreciated figure in the world of statistics in the 1960s she certainly wasn't by the two academics who greatly influenced my career, which has spanned many fields but always with statistics at its heart!

Thank you again for your very interesting paper.

Very kind regards,

Peter Lansley

- - - - - - - -

## References

1. Deeply Trivial, Great minds in statistics: F.N. David versus the patriarchy. August 23 (2017) http://www.deeplytrivial.com/2017/08/great-minds-in-statistics-fn-david.html Accessed 20 December 2022
2. Laird, N.M.: A conversation with F.N. David, Statistical Science 4(3), 235–246 (1989)
3. Denby, L. and Landwehr, J.: A conversation with Colin L. Mallows, International Statistical Review 81(3), 338–360 (2013)
4. Golbeck A.L.: The road less travelled. . . How Elizabeth L. Scott and F.N. David made an impact, Significance Magazine, pp. 36–38 (June 2021)
5. Management Reality: Peter Lansley, MBE, BSc, MSc, PhD, MCIOB, FRCOT (2023) http://www.managementreality.com/biog_pl1.htm Accessed 30 July 2023

# Ending (1977–∞) 11

*I don't think my job in life is to be infulential... [it is] to ask questions and try to find the answers...*

David, F.N.: in *Statistical Science* conversation article (1989)[1] [1]

## 11.1 After Retirement from UCR

David was influential at UCR, but her time there was difficult. Her close colleague David Barton summed it up with this statement:

> In 1967 David left for California where she lived in Kensington (adjacent to Berkeley). She held the appointment at Riverside as Chairman of the statistics department which she set up and where she was both Professor of Statistics and Professor of Biology, Economics, Statistics and Administration. The sting was in the tail, and administration was on what she spent too much time in the fierce internecine struggles that were endemic in one of the less fashionable of the University of California campuses. She flew in from Berkeley on Sunday nights and fled back on Thursdays. She maintained a close association with Berkeley where she became Research Associate in Biostatistics and Emeritus Professor in 1977 (having retired at this point from Riverside). Also from 1977 to 1990 she maintained her interest in applied statistics by means of her forestry contract [2].

The "less fashionable" Riverside campus was very small, with only about 4,000 students. There wasn't a restaurant scene in Riverside, so David encouraged people to get together for intellectual discourse over bagged lunches. David described the environment there as "difficult" when it came to easy conversational exchanges [1].

The "more fashionable" Berkeley campus was always a draw for David. The scene was filled with easy conversational exchanges and rich with intellectual discourse. David left for Berkeley in 1967 and didn't look back. She began the next chapter of her life after UCR (Fig. 11.1).

The year 1980 marked the 150th anniversary of the independence of the country of Belgium. Two major federal agencies in the United States—the National Endowment for the Arts and the National Endowment for the Humanities—decided to organize celebrations. San Francisco was one of three cities chosen to be a focal point for the celebrations. The theme would be BELGIUM TODAY. The San Francisco planning committee invited UCB to contribute. They could provide lectures, seminars,

[1] Note: This was four years before David's death.


A. L. Golbeck, *Florence Nightingale David*,
https://doi.org/10.1007/978-3-031-77768-4_11

**Fig. 11.1** David, F.N. working in her US Forest Service office (around 1977) [Image credit: Reprinted with the permission of the Institute of Mathematical Statistics]

exhibitions, and programs across the arts. The UCB chancellor immediately saw an opportunity to connect its activities to their International House's 50th Anniversary.[2] The International House's mission is "To foster intercultural respect and understanding, lifelong friendships, and leadership skills for a more just and peaceful world" [3].

David R. Brillinger was the chair of the UCB statistics department at the time, and he brought the Belgium opportunity to the attention of the faculty. Neyman came up with an idea: How about involving David in the celebration? She had already written a historical paper on the famous Belgian scholar Adolphe Quetelet, where she analyzed his early involvement with statistics. She could present a public lecture and perhaps also prepare some materials for public display relating to Belgian scientific achievements.[3] Neyman explained: "While the event of Belgium gaining its independence has many aspects, financial and political, it seems essential that at the time that country was marked by scholarly activities of high international importance. Such activities may be symbolized by Adolphe Quetelet... about 150 years ago Quetelet was one of the leaders in research activities on a broad scale, including not only statistics but also astronomy, meterology, etc." Neyman then told Brillinger that David had already compiled a list of Quetelet's activities and reminded him that she was an expert on the history of statistics and taught the History of statistics course for the department.[4]

Neyman jumped into the correspondence with the determination that it would be appropriate to compensate David for time and effort to prepare the lecture. He told Brillinger that he included 2,000 USD in his US Office of Naval Research project renewal budget for this purpose. Her deliverables would be a lecture and a published article.[5] Brillinger agreed with the idea,[6] and the Belgium

[2] Bowker, A.H.: Letter to University of California-Berkeley deans, directors, department chairs, and administrative officers 2 July 1979. Note: Albert H. Bowker was the University of California—Berkeley chancellor.

[3] Brillinger, D.R.: Handwritten letter to Bowker, A.H. 16 July 1979.

[4] Neyman, J.: Letter to Brillinger, D.R. 9 August 1979.

[5] Neyman, J.: Letter to Brillinger, D.R. 6 August 1979.

[6] Brillinger, D.R.: Handwritten letter to Neyman, J. 6 August 1979.

committee did too.[7] Neyman received his grant renewal from the office of naval research, and it included the funding for David's talk.

David prepared the 29 April 1980 public lecture. She prepared the report based on material she compiled for the lecture. She filed the 25-page report, "Adolphe Quetelet: Prophet of the New Statistics," with the Defense Technical Information Center in May 1980. David wrote this abstract:

> Adolphe Quetelet (1796–1874) was born in Ghent and was eventually a professor at the University there teaching mathematics, astronomy, geology, physics and history of science. He was almost entirely responsible for the Royal Observatory in Brussels being built in 1826, and became its director in 1828. An able mathematician, he was concerned also with the collection of data. In 1829 he drew up the plans for the Belgian census and in 1841 created the Central Statistical Commission. On a visit to London in 1834 he became one of the founders of the London (later Royal) Statistical Society and was the first foreign member elected to the American Statistical Association when it began in 1839.
> He devoted his whole life to showing how probability and statistical techniques could be utilized in the development of scientific research and in government administration. At the meeting of the British Statistical Association in 1841 he listed more than forty topics which he thought should be investigated using statistical methods—including anthropology, criminal justice, meteorology, agriculture, zoology and man.

The last paragraph of David's abstract summarized Quetelet's influence: "[Quetelet] was the first bridge over the gap between theory and practice of statistics and did much to create our modern statistical practices."[8]

David's title at this point was research biostatistician, and her position was located in the UCB school of public health's biomedical and environmental health sciences division. She was added to the UCB payroll (3,225 USD) to analyze some data from the forest service, specifically on "the accumulated Central Sierra Profiling snow gage snowpack profile data."[9] Neyman characterized David's effort as "occasional work."[10]

William Kruskal wrote to David asking her to read and comment on his draft book review of a R.A. Fisher biography. The two agreed that a "truly rounded" evaluation of any scholar would have "to wait 50 years or so," but he saw value in early biographies because they could tap recent memories. David shared with Kruskal her memories about Fisher's "contumacious nonresponse" to her questions and her "recollections of why no one wanted Fisher in a wartime group."[11]

Around 1980, David visited a colleague in Gainesville, FL. She was accompanied by her companion Joyce Townend. After David returned home, her host gave her his two cents, or as he put it, his "tuppence." He told her "everyone was delighted with your visit; you and Joyce have made quite an impression on us all. I hope Miss Townend didn't mind the lapse of formality on occasion in the use of her christian name; she has such an enthusiastic interest in everything that formality goes out of the window. Of course, too, everyone hopes you will come here. That last statement is to be read bearing in mind that I'm still given to English understatement." He added to the letter some information and opinions about potential faculty hires, political races, and offprints. He closed the letter with "Well, whatever the outcome of your visit, I know everyone would want to be included in sending best wishes for the future."[12]

---

7 Brillinger, D.R.: Handwritten letter to David, F.N. 27 September 1979.

8 David, F.N.: Adolphe Quetelet: Prophet of the new statistics, UCB Statistical Laboratory, Office of Naval Research, 28 pages, May 1980.

9 Syme, S.L. and Parlette, G.N.: Personnel Action Form 12 November 1975.

10 Neyman, J.: Letter to Kuhl, L.V. 12 December 1979. Note: Leonard V. Kuhl was University of California—Berkeley College of Letters and Sciences dean.

11 Kruskal, W.: Letter to David, F.N. 6 November 1979.

12 John (last name not given): Letter to David, F.N. n.d.

In 1981, David's old UCL colleague Norman L. Johnson, who was then at the University of North Carolina—Chapel Hill, invited her to contribute entries on Neyman, K Pearson, and E Pearson to Volume V of the "Encyclopedia of Statistical Sciences" which he was co-editing with Samuel Kotz.[13] She agreed to do the entry for K Pearson.[14]

Dartmouth College Professor of Mathematics J. Laurie Snell wanted to use David's "very nice" *Games, Gods and Gambling* book, but found that it was out of print. He asked David's permission to make 16 copies of it so he could use it in his class.[15] David didn't know that the book was out of print and suggested that Snell write to the publisher to ask for the copies of the "G. G. and G." or "G3" book, as she called it. David made the same request herself. She also told Snell that "Maybe if you Xerox the first chapter and wait on events things will work out." She then promised to send Snell copies of other historical papers that she had written.[16,17]

Griffin Publishers sent a letter of apology to David, saying they were upset by her letter. They reported that "Macmillian has been selling at about 100 copies per year. So although we are currently discussing slow sellers with them, there is no doubt that GAMES, GODS and GAMBLING is not one of these. [para] I will let you know the outcome of this unfortunate state of affairs."[18] They also sent a letter to Snell, saying that the "G3" book wasn't out of print, it was just out of stock in the United States where it was being sold under the Macmillan publishing company name. They offered to send Snell 16 copies from the United Kingdom.[19] Snell's students wanted to learn from the book and were willing to pay the postage to get the books from the United Kingdom.[20]

David was "in and out of the U.S. a great deal" in those days.[21] Some of her movements were professional, and others personal. An example of the professional is that she accepted an invitation to give a talk, "My Days at University College," to the Birmingham chapter of the Royal Statistical Society on 7 April 1983.[22] She divulged part of her own history as she accepted the invitation:

> The audience will probably have very little idea of the state of development of statistics at the time of the switch from Karl Pearson to Fisher-Neyman-Pearson in 1933. I propose therefore to go back a little in time—say to the death of Quetelet (1874)—and to sketch very briefly the development of statistics up to 1933. I will mention here Karl Pearson's invention and extensive use of supposedly "modern" techniques such as maximum likelihood. I was a student at U.C.L. from 1931 to 1933 and went to lectures by Karl Pearson, Egon Pearson, Percy Stocks (1889–1974), and G. M. [illigible], among others, so I experienced what was the state of the art at that time. [para] In 1933, when Karl Pearson retired from the Galton Chair, I went with him as his solitary research assistant. We had quarters in the Zoology Department and I went to J.B.S. Haldane's seminars in addition to crossing the quadrangle to listen to Fisher and to Neyman, so I was in the middle of the new things as they developed. I want to contrast the years 1933 to 1938 (when Neyman went to the U.S.) with those pre 1933. When I went back to the Statistics Department in 1935 as Neyman's first and only research assistant life was somewhat onerous.

Relating to the personal, David divulged the address of her house as 5 Stuart's Meadow, Graffham, Petsworth, Sussex (T. M. Graffham 221). She said that she would probably be moving around before

[13] Johnson, N.L.: Letter to David, F.N. 10 July 1981.

[14] David, F.N.: Handwritten note on the Johnson letter.

[15] Snell, L.J.: Letter to David, F.N. 1 December 1981.

[16] David, F.N.: Letter to Snell, J.L. 8 December 1981.

[17] David, F.N.: Letter to Griffin, J. 8 December 1981. Note: James Griffin was at Charles Griffin Publishers in London.

[18] Griffin, J.: Letter to David, F.N. 16 December 1981.

[19] Griffin Publishing Company: Letter to Snell, J.L. 15 December 1981.

[20] Snell, J.L.: Handwritten P.S. on a letter to David, F.N. 4 January 1982.

[21] David, F.N.: Handwritten letter to Gokhale, D.V. 3 November 1982. Note: D.V. Gokhale was University of California—Riverside professor of statistics.

[22] Nevill, A.M.: Handwritten letter to David, F.N. 17 January 1983. Note: A.M. Nevill was in the University of Birmingham department of physical education.

the day of her talk, including a stay at Tettenhall (Wolverhampton), so she would probably come to the talk from there rather than her house in Graffham.[23]

Here are some excerpts of David's talk. We are repeating them here, because they articulate the state of the field of statistics at David's time:

> I went to University College London as a postgraduate student in statistics in 1931 and sat at Karl Pearson's feet for two years. When he retired from the Galton Chair in 1933 and went to quarters in the Zoology Department I accompanied him as his last and only research assistant. There I was lucky enough to strike up a friendship with J.B.S. Haldane—we crossed the quadrangle together to attend R.A. Fisher's lectures and seminars—and I also went to those of Neyman. In 1935 I went back to the Statistics Department, as Neyman's first research assistant and became a faculty member under E.S. Pearson in 1937. I tell you all of this because my impressions of this period are essentially derived from a worm's eye view—if worms can be said to have eyes.[24] I am also I regret to say well known as an iconoclast.
>
> I propose to set the stage by reminding you briefly of the historical development of our subject. Adolphe Quetelet died in Brussels in 1874. He had had a hand in the formation of the London Statistical Society—later the R.S.S.—the first of which met in Brussels in 1853. The English had led the world in the process of recording official statistics. Quetelet brought in the idea that data should be comparable both as regards time and space, but his greatest contribution perhaps was his insistence that data can be collected and analyzed about anything from the time of the flowering of plants to the relation of tides with weather conditions. His insistence that data can and should be collected about everything under the sun led to the collection of data becoming a commonplace and made it inevitable that methods of analysis would be developed to deal with it all. Quetelet failed in all but the most primitive methods—one example is his efforts to fit a normal curve using quartiles—and it was not until the advent of Karl Pearson that things really began to take off.
>
> What Karl Pearson taught his students in 1931 was a distillation of 50 years of progress in beating out the analytical tools required by the biological data collectors. Galton and Weldon were his chief coadjutors during the late eighteen hundreds, which possibly was responsible for the biological bias, but by 1900, a great deal had been accomplished, most of which I learnt. All the tools which we now use in one guise or another had been set out. Correlation and regression, Chi Squared contingency tables, generalisations of the binomial, contours of the normal surface, and the repetition over and over again that correlation does not imply causation. The messy business of using quantiles—i.e., quartiles, medians, etc.—was replaced during that time; the Pearson system of frequency curves and the various uses of moments, and the beginnings of maximum likelihood as a tool were all in use before the turn of the century. Pearson's 1894 paper on correlation is worth looking at for a brief moment because it shows where the "revolutionaries" of the nineteen-thirties got some of their ideas.

David then turned her attention to answering the question: "What is the best formula to express the correlation" which she answered using mathematics. She continued:

> Throughout the work of this period and to a certain extent in his lectures later there was an understandable confusion between population and sample values—understandable in that all the data to be dealt with were very large numerically. Our science was developing and the small sample era had not yet arrived. Pearson reacted perhaps too strongly against the general assumption - originating from Quetelet—that all data are normally distributed but he was insistent both in his writings and his teaching that data are rarely normally distributed and the practical work for his classes would consist for about 5 weeks of sitting down with a hand Brunsviga and a mass of frequency distributions and seeing at first hand that the dictum "All data are normally distributed but some are more normal than others" was just not so. Pearson developed ChiSquare in order—among other things—to test the goodness of fit of his curves, but he never accepted and never taught the loss of degrees of freedom which occurs when the parameters of the distributions have to be estimated. Needless to say this estimation was always carried out using the method of moments—a method which, if I had to fit a curve and only had a hand Brunsviga, I would use today. During the late Eighties Pearson and Galton and Weldon and many others developed the law of Ancestral Inheritance—now recently coming back into popularity—and many statistical tools were required and invented and passed into general use. So in my day I learnt contingency tables and multiple correlation and regression, the correlation of ranks and tetrachoric r—in fact I think I learnt all the tools that we use today although we may put a different use and interpretation on some of them...

That is the end of the part of the speech that survived.[25]

---

[23] David, F.N.: Handwritten letter to Neville, A.M. 7 February 1983.

[24] Note: David said she had K Pearson's "sole attention for 2 years. I met almost everyone of importance in the statistical world because when they came to London they sought [K Pearson] out." [1].

[25] David, F.N.: Speech given at Birmingham 5 April 1983.

## 11.2 Still in Demand

A group in New Zealand thought David might be willing and able to write an article about Florence Nightingale, probably given David's track record of writing history of statistics and given that the two shared names. It would be included in a book being developed under the title, *The Fascination of Statistics*.[26,27] David did write six handwritten pages found in her files which looked like lecture notes.[28]

On the first page of her notes, David wrote that Florence Nightingale used to write down her thoughts, and many of these writings somehow survived her. One dated 7 February 1837 stated that "God spoke to me and called me to his service." David noted the reference to Joan of Arc and commented: "[Florence Nightingale] was not informed as to what the service would be and spent some years wondering what."

On the last page of her notes, David wrote "[Florence Nightingale] called herself a Passionate Statistician and Statistics were to her a religious exercise. 'The true foundation of theology is to ascertain the character of God... It is by the art of Statistics that law in the social sphere can be ascertained and codified, and certain aspects of the character of God thereby revealed. The study of statistics is thus a religious service.'"

David went on to tell: "[Florence Nightingale] tried in company with Francis Galton and other friends to give money to found a Professorship of Applied Statistics at Oxford. But there was opposition and after a time the scheme was laid aside in favor of more pressing events."[29]

David noted to the New Zealanders, who were requesting that she write about Florence Nightingale, that Nightingale was influenced by William Farr (b.1807–d.1883). At UCB, David gave a series of history seminars. One was on William Farr, and another was on Florence Nightingale. David continued: "I also, on occasion, lecture about [Florence Nightingale] to Women's Lib. and the like." She then offered the New Zealanders her lecture notes, but pointed out that the notes could only be read with difficulty. She also offered that she could write up her lecture notes for an article, but this would take some time because she would need to check references and add more statistical information.[30] David also offered to cut some anecdotes, but the New Zealanders asked her to leave them in: "... they are delightful and I am sure that our readers would be interested in them."[31]

The book chapter didn't happen. The deadline, even when some flexibility was added, was just too short. The editors were so disappointed that they offered to edit a submission from her. They thought her article would have been one of the most popular in the book.[32]

David was an in-demand instructor. She said she didn't like teaching, but the students liked it and flocked to her classes. Then they remembered her kindly after they graduated, even in retirement. For example, J.C. Koop at North Carolina was retired when he wrote this to David: "My mother always said (in Burmese), honour your teachers and parents. In this statement teachers come before parents. I hope it will not embarrass you, if at this late period of time, I say that your lectures in probability and statistics have helped me a lot in my understanding of the probabilistic basis of sample surveys

[26] Brook, R.J.: Letter to David, F.N. 27 January 1983.

[27] Brook, R.J.: Letter to David, F.N. 21 February 1983.

[28] David, F.N.: Handwritten notes titled "Florence Nightingale (1820–1910)."

[29] David, F.N.: Handwritten pages about Florence Nightingale, n.d.

[30] David, F.N.: Handwritten letter to Brook, R.J. 9 February 1983.

[31] Brook, R.J.: Letter to David, F.N. 21 February 1983.

[32] Brook, R.J.: Letter to David, F.N. 21 March 1983.

and its direct connection with randomization and inference. I thank you and remain deeply grateful for passing on this knowledge." He also chastised David for saying she would not publish about any statisticians until they are dead.[33]

David was in the United States on an immigrant visa, having entered the country on 27 May 1960. In addition to her research biostatistician position at UCB, she held the title of professor emeritus. In 1982 while she was at UCB, she was elected to be a fellow of the American Association for the Advancement of Science.

In 1984, David was still getting requests about her G3 book. One, for example, came from mathematician Flemming Topsoe at Copenhagen University. He said he was "enthusiastic" about her book and wanted to include some of it in an elementary textbook that he was writing. He was especially interested in her discussions of combinatorics.[34]

UCB Statistics Professor Elizabeth L. Scott was a colleague and close friend of David's. In 1985, the chair of the UCB statistics department, David Freedman, wrote to David in support of Scott's promotion to a special salary scale. His letter was four single-spaced pages, and it was glowing. Here is an excerpt:

> "It seems to me that hers [Scott's] is an instance where the record speaks for itself, and that Elizabeth is deserving of whatever award the University can give distinguished professors.", writes F. M. Mosteller of Harvard. "... Betty Scott is a major, significant research scientist of world class; were she at the University of Chicago, I would be selfishly pleased, and I am sure that she would have recognition in our Chicago manner similar to Berkeley's Special Salary mode of recognition.", so writes W.J. Kruskal. These are two general comments on Elizabeth Scott and her work. This report will review various of her contributions to science, education and society and will end with the strongest recommendation for her advancement to Special Salary status. It would be timely and fitting recognition of her distinguished career.

Freedman then summarized the five letters that were written to recommend Scott for the new status, saying "The warmth and respect that the writers have for Betty is clearly evident." One of the writers, a biologist, said "... when she gets interested in a subject, she tracks down and uses all of the data and tries to make a consistent picture not only of the statistical but the biological analysis... From the point of view of the biological community, she is our dream." Mosteller said "Her work... was most original and certainly took the experts in the field [of cancer] by surprise... [her approach involves] great concern with the basic science for the situation and tremendous energy and perserverance [sic] in carrying a project through."

Later in the letter, Freedman mentioned Scott as being responsible for Evans Hall on the Berkeley campus: "As David Blackwell has remarked, 'People think she owns it.' Freedman also talked about Scott's national service, and her service to the department which included her high esteem by the students, concluding Her unselfishness of time, energy and resources is manifest."[35]

Scott in 1988 was nominated for the Berkeley Citation, and David kept a copy of the letter. The Citation is an award given to a person or organization whose contributions both to UCB and academe were outstanding. UCB Professor of Statistics David R. Brillinger wrote the letter, saying that Scott "has had a long and highly influential career on the Berkeley campus and in the scientific world... The Citation would be a timely recognition of her distinguished and energetic efforts... 'The bear does not quit. The bear does not die.' I have always thought that this cry of Joe Kapp's was epitomized by Elizabeth Scott. Let's recognize her Berkeley contributions formally."[36]

---

[33] Koop, J.C.: Handwritten letter to David, F.N. 10 June 1983.

[34] Topsoe, F.: Letter to David, F.N. 30 May 1984.

[35] Freedman, D.A.: Letter to David, F.N. 24 September 1985.

[36] Brillinger, D.R.: Letter to Heyman, M. 19 August 1988. Note: I.M. Heyman was UC chancellor.

Harvard University Biostatistics Professor Nan Laird interviewed David for a *Statistical Science* conversation article which was published in 1989. According to Laird, David wasn't much interested in discussing, or didn't have much memory about, her old research papers. Laird had a lot of extra material from the interview that she didn't include in the article. She commented: "... I'm reluctant to let it out, since [David] was very upset with some of the material I included in an early version... I think [David] probably wants to be helpful... but she does not have much patience or energy."[37] When reprints of the *Statistical Science* article were produced, Laird offered to send some to David, saying "I have had several nice letters from people saying how much they appreciated the article. I hope it was not too painful for you."[38]

Elizabeth L. Scott died in 1988. Royal Statistical Society Honorary Secretary Denise Lievesley asked David if she would write a short obituary of Scott to include in the RSS Journal Series A.[39] David wrote a page by hand, and it was published.

At this point, David was describing herself as "old and decrepit." She had the time to look way back into her days at UCL working as a PhD degree student under K Pearson for 2 years and then following him upon his retirement to be his sole research assistant. Since then, she "had been plagued with queries about K.P.'s early days at U.C.L." In particular, she had been asked about Lucy Whittaker, who was associated with the K Pearson's Biometric Laboratory at UCL. Whittaker had raised some questions about a paper by Gosset having to do with the practical aspects of Poisson's law [4]. David knew there were five letters from Whittaker in the archive, and she requested xerox copies of them for a history project that she was considering doing.[40]

Joan Fisher Box (a daughter of R.A. Fisher and wife of George Box) wrote a biography of Fisher [5]. William Kruskal sent David a draft copy and asked her to comment on it. In her reply, she said what she thought about biographies of people who are still alive or have not been dead for very long:

> I wonder if the book is sufficiently important to warrant the attention which you have given it. I know that the modern tendency is to write biographies of people during their life-time and immediately after their deaths, but I am strongly of the opinion that such efforts are of little value. The place of a scientist (or any other body) in his chosen field cannot be judged correctly until enough time has passed for the biographer to be able to stand aside and see the whole sweep of events—say fifty or one hundred years from now for Fisher. I am aware that by so doing one (probably) loses the gossip side of things, such as whether they liked eggs for breakfast or some such, but this I feel could be left out anyway.
>
> I thought Box, and you yourself, do Fisher too much honor. [I suppose I should put here in parentheses that I am by nature an iconoclast.] I have the feeling that when the historian of the future is able to stand away and look at the development of the mathematics of statistics as a tapestry coming from before Laplace and Gauss and continuing through the 19th century, Fisher will take his rightful but not preeminent place. To illustrate this I enclose a copy of an address I gave at a Biometrics Conference (Pages 7–10 are maximum likelihood.) I don't know where Pearson got his ideas from—possibly Gauss?—but it is easy to see where Fisher got his.
>
> I feel that this enormous battery of small sample theory and the concomitant mathematical manipulation of the normal surface have not added greatly to the prime business of the statistician which is making common sense of the data of experience. Fisher's contribution to statistics—as you so rightly remark—is his development of Quetelet's (and others) ideas as regards the application of the random event. If one followed through Fisher's method of attack one would say, for example, here is a set of data. What are the minimum assumptions other than randomness that I need to feed in in order to get an answer? Instead we feed in a whole paraphernalia of assumptions and conveniently forget that our answer is dependent on having made these. It is difficult at this point in time to assess the importance of Fisher's randomisation [sic] approach or how much he got from his predecessors, but the years will make this clear, and I have the intuition that this is what he will be remembered for. I was amused at your remark about infant prodigies in mathematics. Until Euclid went out of fashion it

[37] Laird, N.M.: Letter to Applebaugh, G.N. 4 January 1989. Note: Gwendolyn Neul Applebaugh was a University of Wisconsin-Eau Claire mathematician.

[38] Laird, N.M.: Letter to David, F.N. 13 October 1989.

[39] Lievesley, D.: Letter to David, F.N. 14 April 1989.

[40] David, F.N.: Handwritten letter to University College London archivist 28 April 1989.

was customary to relate in biographies that the biographee had rediscovered Euclid's propositions at an early age, so that any role regarding mathematical precosity is somewhat suspect. On the whole Fisher was not a great mathematician although he was very good. I remember P. L. Hsu "On the distribution of roots of certain determinantal equations" (Ann. Eugen. 9.250). Hsu and I were colleagues at that time and we went to Fisher's weekly seminar where he gave the essence of his paper "The sampling distribution of some statistics obtained from non-linear equations." (Ann. Eugen. 9.238) Hsu spotted that Fisher's approach was going to lead to an intractable situation so he went home, worked all night, and produced his own method. I remember Fisher being angry—I think because he had difficulty in understanding Hsu's mathematics—and it was some time before he followed what was going on and peace was restored.
Incidentally Fisher, like Pearson and many others, found it difficult to acknowledge that someone had succeeded and that he had failed. This vanity is a male characteristic, I think, but it possibly accounts for the intense rivalries of the twenties and thirties. Fisher was also anti-female. He had lots of secretaries, computing assistants and so on but they were subordinate to him. If I stood up in a seminar and asked him a question he would ignore me completely and go on as if I hadn't spoken. This happened often enough for me to get my (male) neighbor to ask the question and he was always answered. Looking back over forty years it all seems rather silly. I could wish Box had left out the reference to Fisher and the war years. University College, London, was shut and the staff gone either to war jobs or various universities in the "safe zone." Everyone knew this since the discussions had been going on for over a year. It seemed monstrous that Fisher should decide he was going to work in college, that he should break in and that he should slug the beadle[41] who tried to stop him. He could have done what I believe he eventually did and that is work at Rothamsted. That is where J.B.S. Haldane took his own group. It was probably this reputation for unreasonableness which prevented any war-working groups from inviting Fisher to join them. We did an obligatory seventy hour week and whenever I asked why not Fisher—because I admire him even if I don't like him—I was told that they didn't think they could cope with him. Certainly the attempted break-in episode did not resound to his credit.
I think I have rankled enough—probably too much. I am busy right now, in my spare time, looking at Quetelet. I wonder whether I am far enough away from him in time to see him clearly. As I began by saying, it is difficult to see the threads in the pattern at close range.[42]

David's contributions to the history of statistics remain unparalleled. If only we could ask David to update her assessments, now that more time has passed and she could see the contributions of these statisticians even more clearly.

## 11.3 Leaving Us Behind

On 23 July 1993, David died at her Kensington home from lung cancer at age 83. She had been in the United States for 26 years. She never married or had children. She had served 50 years in her profession, but in her characteristic wry manner, she said she was fond of none of her own work [1].

David's death was reported by her close friend Louise LeCam, the wife of UCB statistician Lucien LeCam. The immediate cause of death was respiratory failure due to lung cancer (she had been diagnosed 6 months earlier). She was also suffering from chronic bronchitis.

With Louise's help, David's body was shipped back to England. Per David's request, she was buried in Graffham Cemetery in Graffham, England, the village where her cottage was located. David must have had great affection for the village of Graffham.[43] Dye packed up David's books and journals to send to the University of Arizona in Tucson.[44]

On one of our archive trips to the United Kingdom, we traveled to Graffham and visited David's grave. Her grave was identified on the directory (4G) as you approached the cemetery (Fig. 11.2).

[41] Note: The beadle that David is referring to is a ceremonial officer of a college [6].

[42] David, F.N.: Handwritten letter to Kruskal, W. n.d.

[43] Florence Nightingale David, Certificate of Death, State of California, certificate number 39307003234, 23 July 1993.

[44] Dye, J.: Handwritten letter to Baldwin, J. 6 June 1993.

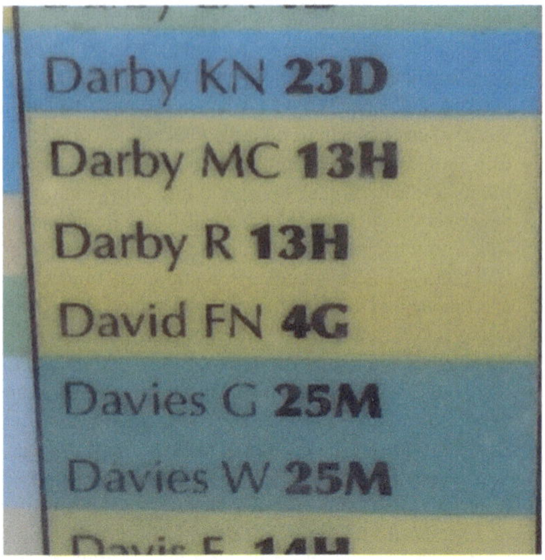


**Fig. 11.2** The grave directory at St. Giles Church in Graffham, England (2019) [Image credit: Golbeck, A.L.]

**Fig. 11.3** The grave where David is buried in Graffham, England, 2019 [Image credit: Golbeck, A.L.]

There was no stone on the grave 4G. We accordingly reported that David was buried in an unmarked grave [7].

Later we got in touch with Rev'd Vivien, the minister of the St. Giles Church in Graffham, asking about how we could purchase a gravestone for David's grave. Rev'd Vivien went to the gravesite and noticed that David was not buried in the grave identified on the directory but rather in the grave next to it along with the remains of her companion Joyce Townend. When we visited the cemetery, we hadn't seen David's name on the Townend stone because there were irises covering the lettering of David's name (Fig. 11.3).

David's *New York Times* obituary mentioned her many publications, including over 100 articles and 9 books. It mentioned that David was survived by a sister in England and a sister in Canada [8].[45] The obit made special note of her history book, *Games, Gods and Gambling,* and the original dust cover with the scantily clad Greek goddess.

A book identified David as one of four women "associated with early history of statistics."[46] The editors called her a "pioneer feminist statistician." They said that "David's low key and humorous approach to biases against women who chose Statistics as a profession actually opened the door to women preferring that scientific area" [9].

An obituary written by an unidentified author who was one of David's friends in the United States[47] said this about her scholarly contributions: "[They] were very varied. In over one hundred scholarly papers she dealt with abstract problems in combinatories as well as enumeration of birds in a forest setting and the spread of forest fires... [para] One of her books on the history of probabilities, entitled *Gods, Games and Gambling*, attracted the ire of the U.S. Customs because of a dust jacket that featured a sketch of a goddess of fortune, scantily clad as befits Greek goddesses. It was allowed into the U.S. only after the U.S. Customs dutifully and painfully removed the dust jackets." The obituary concluded with this: "She will be remembered by her friends as a person of uncompromising integrity with a sharp wit and a fondness for Cuban cigars. No services are being planned. Burial will take place in England. Donations in her memory may be sent to the Charity of your choice."[48] David's next of kin were two sisters, but UCL did not have their contact information.[49]

The same obituary said that David eventually succeeded "...Maurice Bartlett as Head of the UCL Department of Statistics." But this is not accurate. Dennis Lindley succeeded Bartlett, and David never became head of the UCL department of statistics in the United Kingdom. Instead she hit a glass ceiling at UCL and became head of the UCR department of statistics in the United States. Apparently based on this obituary, UCL Director of Administrative Services Huggins published an announcement of David's death which incorrectly identified her as head of department of statistics 1962–1967.[50]

In 2001, the Committee of Presidents of Statistical Societies (COPSS), in partnership with the Caucus for Women in Statistics, established the Florence Nightingale David Award. It is given every 2 years to a woman who has advanced the discipline of statistics and is an outstanding role model. David was given the award retroactively: "For her efforts in opening the door to women in statistics; for contributions to the profession over many years; for contributions to education, science, and public service; for research contributions to combinatorics, statistical methods, applications, and understanding history; and her spirit as a lecturer and as a role model" [10] (Fig. 11.4).

In 2018, COPSS established the Elizabeth L. Scott Lectureship and the F.N. David Lectureship. The lecture named after David was the first ever given at the Joint Statistical Meetings that was named after a woman [11]. The significance of this "first" was explained by the author of this book, who noted that previously many thousands of statisticians were deprived of having the presence of a strong woman intellectual role model at these important meetings for almost 200 years [11].

Richard Waterman Scott knew David through his aunt Elizabeth L. Scott. He had this to say about David:

---

[45] Note: We were unsuccessful at locating the families of either of David's sisters.

[46] Note: The other three were Elizabeth L. Scott, Gertrude Cox, and Florence Nightingale.

[47] Jones, J.R.: Letter to Huggins, A. 29 July 1993. Note: J.R. Jones was at Johnson and Clarence Solicitors, and A. Huggins was the University College London director of administrative services.

[48] Author not named: David, F.N. 23 August 1909, journal not named, n.d.

[49] University College London File Note 2 August1993.

[50] Huggins, A.K.: University College London Notice 30 July 1993.

**Fig. 11.4** Golbeck, A.L.: Speaking at the first David Lecture (2019) [Image credit: Molgaard, C.A.]

> David was a trendsetter in the field and pushed for equality. She was clearly counter-culture on two continents...choosing to live with another female, when this was not overtly accepted. In many ways, she reminded me of Hemingway, although I never met him, but only as he has been portrayed, and as in the books he has written. The times I met her, I knew nothing of her personal life, of course, but I knew she was a colorful person, preferring to dress in more masculine outfits (pants and shirt) and always drew an audience at the parties or social gatherings at the house on 34 Tunnel Road [Elizabeth L. Scott's house] or at the office on campus.[51]

David was truly an unforgettable passionate statistician, probabilist, historian, and leader.

Barton's obituary pointed out that David's career happened in four distinct phases: UCL 1931–1939, World War II work 1939–1946, UCL 1946–1967, and UCR/UCB 1967–1993 [2]. He summed up David's life with these words:

> Those who worked with her and indeed her students at all levels will remember a woman of great generosity and charm. She had a very short fuse and could be extremely rude when she chose. Her enormous energy, prolific output and uncertain temper made her an exciting colleague to work with...her later forestry studies...had military implications... Her contribution to statistics was wide-ranging and substantial.

It also made her an exciting subject for this book about what it was like to be a woman professor in the twentieth century.

## References


1. Laird, N.M.: A conversation with F.N. David, Statistical Science 4(3), 235–246 (1989)
2. Barton, D.E.: Obituary: Florence Nightingale David, Journal of the Royal Statistical Society Series A 157(2), 299–303 (1994)
3. International House at Berkeley (2023) https://ihouse.berkeley.edu/ Accessed 7 June 2023
4. Pearson, E.S.: "Student" as statistician, Biometrika 30(3/4), 210–250 January (1939) https://gwern.net/doc/statistics/decision/1939-pearson.pdf Accessed 3 February 2023


[51] Scott, R.W.: email communication to Golbeck, A.L. 3 October 2022.

5. Box, J.F.: R.A. Fisher: The Life of a Scientist, Wiley (1978)
6. Harte, N., North, J., and Brewis, G.: The World of UCL, Fourth Edition, UCL Press, p. 203 (2018)
7. Golbeck A.L. and Molgaard C.A.: The inimitable F.N. David: A Renaissance statistician, Notices of the American Mathematical Society 68(3), 321–330 (2021)
8. The Associated Press: Dr. Florence N. David, Statistician, Dies at 83. The New York Times 28 July (1993) https://www.nytimes.com/1993/07/28/obituaries/dr-florence-n-david-statistician-dies-at-83.html Accessed 23 June 2024
9. Williams, R.H., Zimmerman, D.W., Ross, D.C. and Zumbo, B.D.: Twelve British Statisticians. eBook, Boson Books, pp. 33–37 (2009)
10. Golbeck, A.L.: Supporting an inclusive community: A Caucus for Women in Statistics, Significance Magazine 17(2), 42–44 (2020)
11. Golbeck, A.L: Statistics association presidents establish Elizabeth L. Scott, F.N. David Lectureships, AmStat News, 1 June (2018) https://magazine.amstat.org/blog/2018/06/01/statistics-association-presidents-establish-elizabeth-l-scott-f-n-david-lectureships/ Accessed 19 December 2022

# Part VII

# Over the Lifetime

# 12 Summarizing

*[The F.N. David Award] is granted to a female statistician who serves as a role model to other women by her contributions to the profession through excellence in research, multidisciplinary collaborative groups, statistics education, or service to the professional societies.*

Committee of Presidents of Statistical Societies: Florence Nightingale David Award (2024) [1]

## 12.1 Some of David's Principles

The Committee of Presidents of Statistical Societies advanced Florence Nightingale David as a role model because she was a consummate professional statistician. Here are ten basic principles, based on our experience with David and her story, that she followed to make her a person whose professional behavior should be emulated by others:

1. Turn setbacks into opportunities.
   David applied to become an actuary and was turned down when it was discovered she was a woman. She heard about K Pearson's interest in actuarial science and knocked on his door. She was respectful to him, and he saw something in her. By refusing to accept defeat, taking a chance, and being respectful, she got a start on an exemplary career in statistics.
2. Get on with your work.
   David's schoolteacher father told her she would face discrimination all her life and she should just get on with her work. That's what she did. She established professional goals based on what she wanted to achieve, not on what she thought might be achievable. She became one of the first women in science to achieve the rank of professor in a research university.
3. Establish worthy priorities early.
   David willingly contributed to the World War II effort in the United Kingdom and the academic women's movement in the United States. Yet, after she was elected to serve as a governor of her alma mater, she only attended one meeting. David's colleague Jerzy Neyman used to call her "Duchess," because she knew how to prioritize work requests according to her values and then say "no."


A. L. Golbeck, *Florence Nightingale David*,
https://doi.org/10.1007/978-3-031-77768-4_12

4. Add breadth when and where you can.
   David followed in the footsteps of her first mentor, the "Renaissance man" K Pearson of UCL. While others in her department were statisticians, David was both a probabilist and statistician (theoretical and applied). She wrote history of probability and was a book review editor. Many academics were viewed as narrow specialists compared with David.
5. Practice lifelong leadership.
   David was the head girl in a highly selective grammar school. She eventually became one of the first women in science to attain a department chair position in a research university. Leadership is influence, but David didn't see that as her job. Nonetheless, she practiced leadership in every position she held, even when she was not a titled administrator.
6. Publish early and often.
   David published her first paper at age 23 (with K Pearson) and her first book at age 29. Altogether, she published nine books with quality publishers—four with Cambridge University Press—and over 100 papers in quality journals. She was repeatedly described as prolific. David was creative and always had a future paper or book in mind.
7. Establish your work ethic.
   Beginning in graduate school, David established an admirable work ethic and pattern of diligence. David used a mechanical calculator to produce her fundamental probability integral tables. Over the course of her career, she cranked the handle of a mechanical calculator over 2 million times. It needed to be done to advance the science.
8. Expand your assignments.
   As a young woman, when David was given a calculation assignment, she took the initiative to expand the parameters of her work to include the theory. She regularly did more than was expected of her, even as much as 200 percent. She also became an acclaimed teacher. David exceeded the norm in the workplace and earned many promotions.
9. Make cultural adaptations.
   David wore suits to work at UCL, but at the University of California, she often wore work shirts, jeans, and heavy work boots. She made breakfast in bed for friends staying overnight but resisted exercising hospitality roles at the office. David did what she deemed necessary to adapt to the cultures of academe in the locations she worked.
10. Find and refine a philosophy.
    David formed her own opinions about her profession. She eventually decided that hypothesis testing was overdone, normal curves should be tossed out the window, and there was too much math trivia being published. She thought data should drive inquiry and statistics should be used to solve real-world problems.

## 12.2 Professional Family Trees

K Pearson's professional family tree is given by Neurotree at https://neurotree.org/beta/tree.php?pid=4213.[1]

The tree indicates that K Pearson was a doctoral student of Francis Galton of Cambridge. It also indicates that David was a doctoral student of K Pearson of UCL,[2] along with ten other luminaries like William Sealy Gosset (a.k.a. *Student*):

[1] Accessed 3 March 2023.

[2] Note: We also note that it was Jerzy Neyman who was her mentor and motivated her to complete her PhD.

- Irwin, Joseph
- Elderton, William
- Gosset, William
- Wechsler, David
- Wishart, John
- Laski, Harold
- Hall, Philip
- Carr-Saunders, Alexander
- David, Florence Nightingale
- Georgescu-Roegen, Nicholas
- Greenwood, Major

David has more identified doctoral students (nine, having supervised three at UCL, and six at UCR) than any of K Pearson's other students.[3]

David's professional family tree is also given by Mathematical Genealogy at https://genealogy.math.ndsu.nodak.edu/id.php?id=49989.[4]

The tree indicates the year of completion for David's nine doctoral students:

- Mallows, Colin (UCL, 1953)
- Jenkins, Gwilym (UCL,1956)
- Saw, John (UCL, 1959)
- Deekshit, Arun (UCR, 1974)
- Maxwell, Floyd (UCR, 1974)
- Runnestrand, Roberta (UCR, 1974)
- Harms, Richard (UCR, 1975)
- Baldwin, James (UCR, 1977)
- Paysen, Timothy (UCR, 1978)

The tree also indicates that David had a total of 37 identified descendants.

## 12.3 Curriculum Vitae

We found a copy of David's "Biography for Academic Personnel." It was not dated, but as far as we can tell, it was from around 1979. Her publications list stopped at 1972.

David had an extensive academic biography, a.k.a., curriculum vitae or CV.

David held various editorial positions, including associate editor of *Biometrika* (1956–1969), editor (UK) of the *International Journal of Abstracts of Statistical Theory and Methods* (1959–1964), book reviews editor for *Biometrics* (1972–1977), referee for papers for various scientific journals, and continuing member of the group in the National Research Council which judged applications for army basic research grants.

David indicated that she was a postgraduate student under Professor K Pearson in 1931–1933. She had a continuation of the Devon Major Exhibition during that period for research under him.[5]

She prepared this summary of her employment history:

---

[3] Note: Laski has the next largest number of identified students, with four.

[4] Accessed 23 June 2024.

[5] David, F.N.: University College London Academic Staff report (1940).

| Dates | Institution | Rank, title, or position |
|---|---|---|
| 1933–1935 | UCL Department of Statistics | Research Assistant to Karl Pearson |
| 1935–1939 | UCL Department of Statistics | Lecturer |
| 1939–1940 | Ordinance Board, Ministry of Supply, UK | Sr. Statistician and Experiment Officer |
| 1940–1945 | Research and Experiments Dept., Ministry of Home Security, UK | Senior Statistician |
| 1945–1947 | Experimental Establishment and D.D. Royal Engineer Equipment | Scientific Advisor |
| 1946 | UCB summer session | Lecturer |
| 1947–1948 | Beach Clearance Committee, Combined Operations H.W. | Scientific Advisor |
| 1948–1956 | Board of Studies in Statistics | Secretary |
| 1945–1967 | University College, London | Lecturer, reader, professor of probability and statistics |
| 1948–1966 | UCB summer session | Visiting Prof., Research Statistics |
| 1961–1962 | USAF, Applied Climatology Div. | Research Statistician V |
| 1967–1977 | University of California, Riverside, California | Professor of Statistics I: VI, and overscale |
| 1967–1977 | University of California, Riverside, California | Professor of Biology, Economics, Statistics, and Administration |
| 1977–pres. | University of California, Berkeley, California | Research Biostatistician and Professor Emeritus |
| 1977–pres. | USDA (US Forestry Commission) | Temporary Executive |
| 1978–1979 | University of London, Queen Mary College | Visiting Professor |

David noted that the USDA position was limited to 100 days in any one fiscal year.
David added the following supplementary employment data:

| Dates | Institution | Rank, title, or position |
|---|---|---|
| 1963–pres. | USDA Federal Forestry Service | Consultant |
| 1965–pres. | Pacific State Hospital, Pomona | Consultant |
| 1968–1977 | Dept. of Statistics, UCR | Chairman |
| 1969–1977 | Experiment Station | Associate |
| 1970–1977 | Dept. of Economics, UCR | Professor of Statistics |
| 1970–1977 | Dept. of Biology, UCR | Professor of Statistics |
| 1970–1977 | School of Administration, UCR | Professor of Statistics |
| 1978–1979 | University of Warrick, London | Seminar Leader |
| 1978–1979 | School of Economics, University of Oxford | Seminar Leader |
| 1978–1979 | Polish National Academy Institute of Mathematics | Seminar Leader |
| 1978–1979 | University of California, Berkeley | Seminar Leader |

She summarized her education as follows:

| Years | Name of school | Location | Major subject | Degrees | Date conferred |
|---|---|---|---|---|---|
| 1920–1928 | several schools | United Kingdom | – | – | – |
| 1928–1931 | University of London, Bedford College | United Kingdom | Math | BSc | 1931 |
| 1931–1933 | University of London, University College | United Kingdom | Math/Stat | PhD | 1938 |
| – | University of London, University College | United Kingdom | Statistics | DSc | 1952 |

David listed the following memberships:

- American Statistical Association
- Biometrics Society
- Institute of Mathematical Statistics
- International Statistical Institute
- Royal Statistical Society
- Member, National Research Council (Army Basic Research) 1977–present
- Member, Senior Common Room, Jesus College Oxford, 1977–1978

She listed the following honors and awards:

- Elected Fellow, Institute of Math. Stat (1946)
- Recognized as a Teacher at the Univ. of London (1947)
- Elected as a nonprofessional member of the Professional Board (1947)
- DSc given for "Major contribution" to subject (1952)
- Elected member International Statistical Institute (restricted to 420 worldwide) 1952
- Elected fellow, Amer. Statist. Assoc. (1954)
- Elected Member of the Univ. Senate, Academic Council, Board of Faculty of Science (1955)
- Appointed to the Established Readership in Probability and Statistics (1955)
- "Governor"—Bedford College of Women, Univ. of London (1956)
- Assoc. Editor, Biometrika (1956–1969)
- Editor (United Kingdom) of International J. of Abstracts of Statistical Theory and Methods (1959–1964)
- Distinguished Teaching Award, Univ. of Calif. Riverside (1970–1971)

Finally, David listed this bibliography:[6]
**Books**


1. David, F.N.: Tables of the Correlation Coefficient, Biometrika Trust (2009 Cambridge University Press edition, 64 pages) (1938)
2. David, F.N.: Probability Theory for Statistical Methods (1951 Cambridge University Press Second Edition, 230 pages) (1949)
3. David, F.N.: Elementary Statistical Exercises, Department of Statistics, University College London (1961 Cambridge University Press edition, with E. S. Pearson, 108 pages) (1953)
4. David, F.N.: A Statistical Primer, London: Charles Griffin and Co. Ltd., 226 pages (Japanese Edition: 1961) (1953)
5. David, F.N. and Barton, D.E.: Combinatorial Chance, London: Charles Griffin and Co. Ltd. (1962 Hafner Publishing Company Edition, 356 pages) (1962)
6. David, F.N.: Games, Gods and Gambling: A History of Probability and Statistical Ideas, Charles Griffin and Co. Ltd. (1998 Dover publications edition, 275 pages) (1962)
7. David, F.N.: Research Papers in Statistics: Festschrift for J. Neyman, John Wiley and Sons, 468 pages (1966)
8. David, F.N., Kendall, M.G. and Barton, D.E.: (prepared at the request of Biometrika). Symmetric Function and Allied Tables, Cambridge University Press, 278 pages (1966)


[6] We have edited the references so that they are more complete and in sequential order by year of publication.

9. David, F.N.: A First Course in Statistics, Griffin Monographs No. 31, 228 pages (revised, first published in 1953 as A Statistical Primer) (1971)
10. David, F.N. and Barton, D.E.: Normal Centroids, Medians and Scores for Ordinal Data (Tracts for Computers) Cambridge University Press, 201 pages (1968)

**Monographs**

11. David, F.N.: The Statistical Analysis of Mass Casualty Data, M.O.H.S. 50 pages (1945)
12. David, F.N.: The Blast Actuation of Anti-Tank and Anti-Personnel Mines, M.O.H.S. 156 pages (1945)

**Papers**

13. Pearson, K., Stouffer, S.A., and David, F.N.: Further applications in statistics of the $T_m$ (x) Bessel Function. Biometrika 24(3–4), 293–350 (1932)
14. David, F.N.: On the *p* test for randomness: Remarks, further illustration, and table... Biometrika 26(1–2), 1–11 (1934)
15. David, F.N.: A note on unbiased limits for the correlation coefficient. Biometrika 29, 157–160 (1937)
16. David, F.N.: Limiting distributions connected with certain methods of sampling human populations. Statistical Research Memoirs I (Neyman, J. and Pearson, E.S., eds.) University of London, pp. 69–90 (1938)
17. David, F.N. and Neyman, J.: An extension of the Markoff theorem on least squares. Statistical Research Memoirs, (Neyman, J. and Pearson, E.S., eds.) University of London II, pp. 105–116 (1938)
18. David, F.N.: On Neyman's "Smooth" Test for Goodness of Fit, Biometrika 31(1–2), 191–199 (1939)
19. David, F.N. and Garwood, F.: Applications of mathematical statistics to the assessment of the efficacy of bombing. (i) Theoretical assessment of material damage in random bombing. Civil Defense Research Committee. N. 338 (1942)
20. David, F.N. and Garwood, F.: Applications of mathematical statistics to the assessment of the efficacy of bombing. (ii) Theoretical assessment of Systematic bombing. Civil Defense Research Committee. N. 339 (1942)
21. David, F.N. and Garwood, F.: Applications of mathematical statistics to the assessment of the efficacy of bombing. (iii) The contamination of H.E. craters by gas bombs. Civil Defense Research Committee. N. 340 (1942)
22. David, F.N.: A $\chi^2$ "smooth" test for goodness of fit. Biometrika 34 (1947)
23. David, F.N.: A power function for tests of randomness in a sequence of alternatives. Biometrika 34(3–4), 335–339 (1947)
24. David, F.N. and Johnson, N.L.: The probability integral transformation when parameters are estimated from the sample. Biometrika. 35(1–2), 182–190 (1948)
25. David, F.N.: Correlations between $\chi^2$ cells. Biometrika 35(3–4), 418–421 (1948)
26. David, F.N.: The moments of the z and F distributions. Biometrika 36(3–4), 394–403 (1949)
27. David, F.N.: Note on the application of Fisher's k-statistics. Biometrika 36(3–4), 383–393 (1949)
28. David, F.N. and Kendall, M.G.: Tables of symmetric functions-Part I. Biometrika 36(3–4), 431–449 (1949)
29. David, F.N.: Two combinatorial test of whether a sample has come from a given population. Biometrika 37(1–2), 97–110 (1950)
30. David, F.N. and Johnson, N.L.: The probability integral transformation when a variable is discontinuous. Biometrika 37(1–2), 42–49 (1950)
31. David, F.N.: An alternative form of $\chi^2$. Biometrika 37(3–4), 448–451 (1950)

32. David, F.N. and Johnson, N.L.: The effect of non-normality on the power function of the f-test in the analysis of variance. Biometrika 38(1–2), 43–57 (1951)
33. David, F.N. and Kendall, M.G.: Tables of symmetric functions parts II and III. Biometrika 38(3–4), 435–462 (1951)
34. David, F.N. and Johnson, N.L.: The sensitivity of analysis of variance tests with respect to random between groups variation. (with N. L. Johnson) Trabajos de Estadistica 2, 179–188 (1951)
35. David, F.N. and Johnson, N.L.: A method of investigating the effect of non-normality and heterogeneity of variance on tests of the general linear hypothesis. Annals of Mathematical Statistics 22(3), 382–392 (1951)
36. David, F.N. and Johnson, N.L.: Extension of a method of investigating the properties of analysis of variance tests to the case of random and mixed models. Annals of Mathematical Statistics 23(4), 495–601 (1952)
37. David, F.N. and Johnson, N.L.: The truncated Poisson. Biometrics 8(4), 275–285 (1952)
38. David, F.N.: A note on the evaluation of the multivariate normal integral. Biometrika 40(3–4), 458–459 (1953)
39. David, F.N. and Johnson, N.L.: A test for skewness with ordered variables. Annals of Eugenics London 18 (1954)
40. David, F.N. and Johnson, N.L.: Statistical treatment of censored data: Part I. Fundamental Formulae. Biometrika 41(1–2), 228–240 (1954)
41. David, F.N. and Kendall, M.G.: Tables of symmetric functions. Part IV. Biometrika 40(3–4), 427–446 (1953)
42. David, F.N. and Moore, P.G.: Notes on contagious distributions in plant populations. Annals of Botany 18(1), 47–53 (1954)
43. David, F.N. and Johnson, N.L.: Tests for skewness and kurtosis with ordered variates. Proceedings of the International Congress of Mathematicians—Amsterdam, II, p. 286 (1954)
44. David, F.N.: Studies in the history of probability and statistics: I. Dicing and gaming (A note on the history of probability). Biometrika 42(1–2), 1–15 (1955)
45. David, F.N.: The transformation of discrete variables. Annals of Human Genetics 19(3), 174–182 (1955)
46. David, F.N. and Kendall, M.G.: Tables of symmetric functions. Part V. Biometrika 42(1–2), 223–242 (1955)
47. Barton D.E and David, F.N.: Tests for randomness of points on a line. Biometrika 43(1–2), 104–112 (1956)
48. Barton, D.E. and David, F.N.: Sums of ordered intervals and distances. Mathematika 2(2), 150–159 (1955)
49. Barton, D E. and David, F.N.: Some notes on ordered random intervals. Journal of the Royal Statistical Society Series B (Methodological) 18(1), 79–94 (1956)
50. David, F.N. and Johnson, N.L.: Some tests of significance with ordered variables. Journal of the Royal Statistical Society Series B (Methodological) 18(1), 1–31 (1956)
51. David, F.N. and Johnson, N.L.: Reciprocal Bernoulli and Poisson variables. Metron 18, 77–81 (1956)
52. David, F.N.: Note on Wilcoxon's and allied tests. Biometrika 43(3–4), 485–488 (1956)
53. Barton, D.E. and David, F.N.: "Spearman's rho" and matching problem. British Journal of Mathematical and Statistical Psychology 9(2), 69–73 (1956)
54. Barton, D.E. and David, F.N.: Multiple runs. Biometrika 44(1–2), 168–178 (1957)
55. David, F.N. and Moore, P.G.: A bivariate test for the clumping of supposedly random individuals. Annals of Botany 21(2), 315–320 (1957)

56. David, F.N.: Mr Newton, Mr Pepys and Dyse: A Historical Note. Annals of Science 13(3), 137–147 (1957)
57. Barton, D.E. and David, F.N.: Runs in a ring. Biometrika 45(3–4), 572–578 (1958)
58. David, F.N.: A test for birth order effect. Annals of Human Genetics (1958)
59. Barton, D.E., David, F.N., and Mallows, C.L. (1958) Non-randomness in a sequence of two alternatives: I. Wilcoxon's and allied test statistics. Biometrika 45(1–2), 166–180 (1958)
60. Barton, D.E. and David, F.N.: Non-randomness in a sequence of alternatives II. Runs Test. Biometrika 45(1–2), 253–256.
61. Barton, D.E. and David, F.N.: Sequential occupancy. Biometrika 46(1–2), 218–223 (1959)
62. Barton, D.E. and David, F.N.: Contagious occupancy. Journal of the Royal Statistical Society Series B (Methodological) 21(1), 120–133 (1959)
63. Barton, D.E. and David, F.N.: The dispersion of a number of species. Journal of the Royal Statistical Society Series B (Methodological) 21(1), 190–194 (1959)
64. Barton, D.E. and David, F.N.: Haemocytometer counts and occupancy theory. Trabajos de Estadistica 10, 13–18 (1959)
65. David, F.N.: (with Barton, D. E.) Models of functional relationship illustrated on astronomical data. Bulletin of the International Statistical Institute (1959)
66. David, F.N.: The z-test and symmetrically distributed random variables. Biometrika 46(1–2), 123–129 (1959)
67. Barton, D.E. and David, F.N.: Combinatorial extreme value distributions. Mathematika 6(1), 63–76 (1959)
68. Barton, D.E. and David, F.N.: A collector's problem. rabajos de Estadistica 10, 75–88 (1959)
69. Barton, D.E., David, F.N., and Fix, E.: The polykays of the natural numbers. Biometrika 47(1–2), 53–59 (1960)
70. Barton, D.E. and David, F.N.: The congregating of worms and woodlice. Trabajos de Estadistica 11, 187–197 (1960)
71. Barton, D.E., David, F.N., and O'Neill, A.F.: Some properties of the distribution of the logarithm of non-central f. Biometrika 47(3–4), 417–431 (1960)
72. Barton, D.E., David, F.N., and Merrington, M.: Tables for the solution of the exponential equation, exp(-a) + ka = 1. Biometrika 47(3–4), 439–445 (1960)
73. Barton, D.E. and David, F.N.: The central sampling moments of the mean in samples from a finite population (Aty's formulae and Madow's central limit). Biometrika 48(1/2), 199–201 (1961)
74. David, F.N. and Mallows, C.L.: The variance of Spearman's rho in normal samples. Biometrika 48(1–2), 19–28 (1961)
75. David, F.N.: History of Statistics. Chambers Encyclopædia (1961)
76. Barton, D.E. and David, F.N.: The analysis of chromosome patterns in the normal cell. Annals of Human Genetics 25, 323–329 (1961)
77. David, F.N.: (with Barton, D.E.) Randomisation bases for multivariate tests. Part I. The Bivariate case. Proceedings of the 32nd Session of the International Statistical Institute (1961)
78. David, F.N.: Pressure as a function of distance, part I. U.S.A.F. Scientific Report. (AFCRL-62–461) (1962)
79. David, F.N.: An effective stochastic model for interpreting the correlation between the precipitation at two stations.U.S.A.F. Scientific Report (AFCRL-62-495) (1962)
80. David, F.N. and Gupta, M.K.: Pressure as a function of distance, part II. U.S.A.F. Scientific Report (AFCRL-62–523) (1962)
81. David, F.N. and Fix, E.: Rank correlation and regression in a non-normal surface. Proceedings of the Fourth Berkeley Symposium on Mathematical Statistics and Probability 1, 177–197 (1961)

82. Barton, D.E., David, F.N., and Fix, E.: Persistence in a chain of multiple events when there is simple dependence. Biometrika 49(3–4), 351–357 (1962)
83. Barton, D.E. and David, F.N.: The analysis of chromosome patterns in the abnormal cell. Annals of Human Genetics 26(4), 347–348 (1963)
84. Barton, D.E., David, F.N., and Merrington, M.: Numerical analysis of chromosome patterns. Annals of Human Genetics 26(4), 349–354 (1963)
85. Barton, D.E., David, F.N., and Fix, E.: Random points in a circle and the analysis of chromosome patterns. Biometrika 50(1–2), 23–29 (1963)
86. Barton, D.E., David, F.N., and Merrington, M.: Table for the solution of the exponential equation $\exp(b) - b/(1-p) = 1$. Biometrika 50(1–2), 169–176 (1963)
87. David, F.N.: Stochastic models in meteorology. U.S.A.F. Scientific Report (1963)
88. Barton, D.E., David, F.N., and Merrington, M.: The position of the sex chromosomes in the human cell in mitosis. Annals of Human Genetics 28(1–3), 123–128 (1964)
89. David, F.N. and Fix, E.: Intersections of random chords of a circle. Biometrika 51(3–4), 373–379 (1964)
90. Barton, D.E., David, F.N., and Merrington, M.: A criterion for testing contagion in time and space. Annals of Human Genetics 29(1), 97–102 (1965)
91. Barton, D.E., David, F.N., and Merrington, M.: The relative positions of the chromosomes of the human cell in mitosis. Annals of Human Genetics 29(2), 139–146 (1965)
92. David, F.N.: Some notes on Laplace. Bernoulli-Bayes-Laplace Anniversary Volume, Berkeley Statistical Laboratory Proceedings, Springer, 30–44 (1965)
93. David, F.N. and Barton, D.E.: Two space-time interaction tests for epidemicity. British Journal of Preventive and Social Medicine 20(1), 44–48 (1966)
94. Barton, D.E. and David, F.N.: A persistence problem in renewal theory: Robert the Bruce's spider. Biometrika 53(1), 255–258 (1966)
95. Barton, D.E. and David, F.N.: The random intersection of two graphs. Festschrift for J. Neyman, pp. 445–460 (1966)
96. David, F.N. and Fix, E.: Randomisation and the serial correlation coefficient. Festschrift for J. Neyman, pp. 461–468 (1966)
97. Barton, D.E., David, F.N., Fix, E., Merrington, M, and Mustacchi, P.: Tests for space-time interaction and a power function. (with Barton, D.E., Fix, E., Merrington, M., and others), Proceedings of the Fifth Berkeley Symposium on Mathematical Statistics and Probability, Volume IV: Biology and problems of health, University of California Press, pp. 217–227 (1965)
98. Barton, D.E., David, F.N., Fix, E., and Merrington, M.: A review of analysis of karyographs of the human cell in mitosis. Proceedings of the Fifth Berkeley Symposium on Mathematical Statistics and Probability, Volume 5, University of California Press, pp. 349–366 (1967)
99. Mustacci, P., David, F.N., and Fix, E.: Three tests for space-time interaction: A comparative evaluation. Proceedings of the Fifth Berkeley Symposium on Mathematical Statistics, Volume 5, University of California Press, pp. 229–215 (1967)
100. Barton, D.E. and David, F.N.: Four letter words: The distribution of pattern frequencies in ring permutations. Journal of the Royal Statistical Society Series B (Methodological) 29(3), 550–569 (1967)
101. David, F.N.: Matched groups and randomisation sets. (with P. J. Kim). Annals of Human Genetics 31, 21–27 (1967) [7]

[7] Note: This reference is as listed on David's Biography for Academic Personnel, but we were unable to locate it in the stated journal.

102. David, F.N.: Combinatorial notation in the nineteenth century. (Contribution to J. O. Irwin's paper on W. A. Whitworth). Journal of the Royal Statistical Society Series A (Statistics in Society), 130, 229–242 (1967)
103. David, F.N.: Francis Galton. Encyclopedia of the Social Sciences. Chainbers: pp. 48–52 (1968)
104. Barton, D.E. and David, F.N.: Sequential occupancy with classification. Biometrika 55(1), 229–241 (1968)
105. David, F.N.: Measurement of diversity I. Proceedings of the Sixth Berkeley Symposium on Mathematical Statistics and Probability, University of California Press, 1, 631–648 (1972)
106. David, F.N.: Measurement of diversity: Multiple Cell Contents. Proceedings of the Sixth Berkeley Symposium on Mathematical Statistics and Probability, University of California Press, 4, 109–136 (1972)
107. David, F.N.: Applications of Neyman's $C(_\alpha)$ technique. Proceedings of the Sixth Berkeley Symposium on Mathematical Statistics, University of California Press, 4, 223–230 (1972)
108. David, F.N.: Symmetric functions, Discrete distributions and diversity. (Paul Dwyer Birthday Volume, Symmetric Functions in Statistics, edited by Tracy, D.S.) (1972)

David's bibliography shows that she published her papers most often in *Biometrika*. Johnson, together with Samuel Kotz, wrote that David "contributed to: distribution theory, combinatorics, statistics and genetics, robustness, history of statistics."[8]

Based on the number of citations, the following papers appear to have been the most referenced:

- David, F.N. and Johnson, N.L.: The probability integral transformation when parameters are estimated from the sample. Biometrika 35(1–2), 182–190 (1948)
- David, F.N. and Johnson, N.L.: The effect of non-normality on the power function of the f-test in the analysis of variance. Biometrika 38(1–2), 43–57 (1951)
- David, F.N. and Johnson, N.L.: The truncated Poisson. Biometrics 8(4), 275–285 (1952)
- David, F.N.: A note on the evaluation of the multivariate normal integral. Biometrika 40(3–4), 458–459 (1953)
- David, F.N. and Johnson, N.L.: Statistical treatment of censored data: Part I. Fundamental Formulae. Biometrika 41(1–2), 228–240 (1954)
- David, F.N. and Moore, P.G.: Notes on contagious distributions in plant populations. Annals of Botany 18(1), 47–53 (1954)
- David, F.N. and Barton, D.E.: Two space-time interaction tests for epidemicity. British Journal of Preventive and Social Medicine 20(1), 44–48 (1966)

Most of these papers were published in the 1950s. The Notes on contagious distributions paper was the most cited, with 445 citations.

David published frequently with D.E. Barton but also often with M.G. Kendall and N. Johnson. Of the approximately ten papers with Johnson on a variety of topics between 1948 and around 1956, he thought the most important was the one presented to the Royal Statistical Society on "dirty tests" with order statistics [2,3]. In addition, David has several papers with F. Garwood, C. Mallows, E. Fix, and M. Merrington. Given how close she was with Neyman, it is interesting that she published only one paper with him. It is unclear whether she used a consistent rule for the order of authorship (order based on the extent of contribution, alphabetic order based on last name, etc.).

---

[8] Johnson, N. L. and Kotz, Samuel. David, Florence Nightingale. IN *Leading Personalities in Statistical Sciences*. Wiley, 1997, pp. 91–93.

David's papers on space-time interaction in epidemic studies, the distribution of chromosomes, and other topics came out of her close collaborations with L.S. Penrose who succeeded R.A. Fisher (who succeeded K Pearson) as Head of Eugenics at UCL. Her papers on statistical ecology came out of her association with the botanists at UCL under Pearsall as head, and these laid the groundwork for more papers on diversity at UCR.

David's scholarly work was important. Laird commented that:

> Many of [David's articles] are actively referred to today. *Combinatorial Chance* (with D.E. Barton) contains fascinating combinatorial probability theory (much like Feller's Volume 1) impossible to find elsewhere. Her Tables of Symmetric Functions contains a 50-page introduction that is still the standard reference. Her *Probability Theory for Statistical Methods* contains the only available treatment of "Lexis Theory," a forerunner of contingency table analysis. Her book, *Games, Gods and Gambling,* is widely recommended as an entertaining and authoritative account of the history of probability [4].

## References

1. Committee of Presidents of Statistical Societies: Florence Nightingale David Award (2024) https://community.amstat.org/copss/awards/fn-david#AboutFNDavid Accessed 23 June 2024
2. Read, C.B.: A conversation with Norman L. Johnson. Statistical Science 19(3), 544–560 (August 2024)
3. David, F.N. and Johnson, N.L.: Some tests of significance with ordered variables. Journal of the Royal Statistical Society Series B (Methodological) 18(1), 1–31 (1956)
4. Laird, N.M.: A conversation with F.N. David, Statistical Science 4(3), 235–246 (1989)

The manufacturer's authorised representative in the EU is Springer
Nature Customer Service Centre GmbH, Europaplatz 3, 69115 Heidelberg,
Germany. If you have any concerns regarding our products, please
contact ProductSafety@springernature.com

Printed and bound by CPI Group (UK) Ltd, Croydon, CR0 4YY
30/09/2026
02225600-0019